A SERIOUS GLANCE
AT CHEMISTRY
Basic Notions Explained

A SERIOUS GLANCE
AT CHEMISTRY
Basic Notions Explained

Milan Trsic
Universidade de São Paulo, Brazil

Evelyn Jeniffer de Lima Toledo
Universidade de São Paulo, Brazil

World Scientific

NEW JERSEY · LONDON · SINGAPORE · BEIJING · SHANGHAI · HONG KONG · TAIPEI · CHENNAI

Published by

Imperial College Press
57 Shelton Street
Covent Garden
London WC2H 9HE

Distributed by

World Scientific Publishing Co. Pte. Ltd.
5 Toh Tuck Link, Singapore 596224
USA office: 27 Warren Street, Suite 401-402, Hackensack, NJ 07601
UK office: 57 Shelton Street, Covent Garden, London WC2H 9HE

British Library Cataloguing-in-Publication Data
A catalogue record for this book is available from the British Library.

A SERIOUS GLANCE AT CHEMISTRY
Basic Notions Explained

ISBN-13 978-1-84816-530-4
ISBN-10 1-84816-530-7

Typeset by Stallion Press
Email: enquiries@stallionpress.com

Printed in Singapore.

I dedicate this book to my son Manuel, who left this world on July 1st. 2008.

Milan Trsic

I dedicate this book to my parents Egnaldo and Sandra.

Evelyn Jeniffer de Lima Toledo

São Carlos, May 2010

Acknowledgments

Let us first express our satisfaction for the confidence we received from Imperial College Press. Our proposition was a chemistry text for people whose main interest is not necessarily chemistry. Nonetheless, those subjects we chose to cover were to be treated at a serious level. We hope that the result offered to the reader fulfills our aim.

Next, we wish to recognise the professionalism of Dr Lee Kok Leong and Mr V. K. Sanjeed, from World Scientific, who were always prompt to follow up and encourage the progress of our work.

We ought to acknowledge that at various instances we searched for opportune information from Google™ and in Wikipedia, the free online encyclopedia.

The quantum chemical calculations reported in the book were performed with the GAUSSIAN03 programme using the machines of the Quantum Chemistry Group of the Instituto de Química de São Carlos, University of São Paulo. All facilities of the Instituto were at our disposal for our work, with full support from the University of São Paulo.

For many years the Brazilian Agencies CNPq, FAPESP, FINEP and CAPES generously supported our Quantum Chemistry Group. One of us (EJLT) has presently a MSc fellowship from CAPES.

Contents

Acknowledgments   vii

Preface   xi

Chapter 1   Introduction   1

Chapter 2   The Electronic Structure of Atoms   3

Chapter 3   The Electronic Structure of Molecules   15

Chapter 4   Organic Molecules   33

Chapter 5   Inorganic Molecules   49

Chapter 6   The Birth of Quantum Mechanics and Modern
            Chemistry   63

Chapter 7   Molecular Orbital Theory and the Recovery of
            Classical Chemical Notions   69

Chapter 8   The Nature of the Atomic Nucleus   91

Chapter 9   Chemical Transformations and Reactions:
            Velocity and Energy Balance   101

Chapter 10 Chemistry and Energy Sources   121

Chapter 11   Chemistry and Living Beings   133

Final Remarks   151

Interview with Professor Rezende (Environmental Chemistry)   157

Notes   163

Further Reading   167

Appendix I   Spherical Co-ordinates   171

Appendix II   Periodic Table of the Elements   173

Appendix III   Constants and Conversion Factors   175

Appendix IV   The Most Stable Ions of the Common   179
               Elements

Appendix V   Isotopes of the Elements and Their Relative   181
              Abundance

Index   185

Preface

When I initiated my appointment in 1978 at the University of São Paulo, one of my first assignments was to teach Chemistry II for physics undergraduate students. It was a very difficult task. For topics such as the electronic structure of atoms, there was a good dialogue with the students; but other subjects, as for instance chemical thermodynamics or kinetics, were of little interest to the audience. I did my best, with the aid of seminars and discussion of articles.

Some 30 years later, I decided to undertake the same challenge with an optimistic approach. I proposed a new programme for Chemistry II for the physics students, which was approved by the Undergraduate Teaching Committee. The strategy was to cover some subjects of the old programme with some of the less appealing subjects approached in a round-about way. I also taught some more chemistry through every-day phenomena such as energy sources, pollution, cracking, chemical structures and processes in living beings.

The semester was a success, the students made very enthusiastic comments at several instances and almost everybody passed with good marks.

In this book, I have left aside the jargon used to attract the attention of physics students. My hope is to make some basic notions of chemistry palatable for readers with main interests in subjects other than chemistry. At the same time, the purpose of the book is to show some examples of how chemistry is strongly related to our experiences, needs, and life itself.

I was halfway through writing the book, when I started to become slightly nervous about the approaching due date for delivery of the manuscript. At that time Evelyn Jeniffer de Lima Toledo became my MSc student. I invited her to join me in the endeavour and she brought much enthusiasm and basic knowledge. I shall add that Jennifer ended her Bachelor degree as co-author of four articles in journals of reasonable circulation. One of these papers contains results from her Graduate Research Assignment.

Milan Trsic
São Carlos, December 2009

CHAPTER 1

Introduction

As the title of this book, *A Serious Glance at Chemistry: Basic Notions Explained,* suggests, we intend to provide some carefully chosen concepts of our present day understanding of chemistry, without subjecting the reader to many mathematical and technical refinements and intricacies. We also, arbitrarily, skip topics which we believe to be of minor interest to readers not oriented to become professional chemists. Whenever possible, we make connections between chemical processes or materials and everyday human needs and concerns. We also emphasise the crucial role of chemistry in living beings.

In Chapter 2, we discuss in some detail the electronic structure of the hydrogen atom. On that basis, many-electron atoms are described qualitatively.

Chapter 3 provides some rationale on how molecules are built, electron density distribution and chemical bonds.

In Chapters 4 and 5, we present organic and inorganic molecules at the level and detail suitable for the purpose of this book.

Modern physico-chemical science is riddled with concepts brought in by quantum mechanics, born in the beginning of the twentieth century. It seems quite appropriate to describe this intellectual creation in Chapter 6 and how it gave origin to a new science, quantum chemistry. The manner in which these new concepts penetrated our concepts of molecular structure is the subject of Chapter 7.

Chapter 8 describes the atomic nucleus as a fundamental piece of the atomic structure; this chapter includes a few remarks on the present notion of the building of elementary particles.

In Chapter 9, we describe various types of chemical transformations and some notions on thermodynamic and kinetic control of chemical reactions.

Chapter 10 is the main piece of this work, focusing on the important issue of energy sources. Special attention is given to chemical-related energy sources, as combustion and fuel cells, although other options for energy production are listed and briefly commented on. We strongly address the risks to our environment and humanity itself.

We consider the closing Chapter 11 as important or more so than the previous chapter. We attempt to summarise most of the structures and processes in which chemistry, biochemistry, biology and biophysics have a crucial role in living beings. We go from relatively simple systems such as vitamins to extremely complex structures such as enzymes and receptors. Always with the aim of avoiding elaborate specific terms and concepts, a marvelous nature is presented to the reader, including, vision, nervous conduction, enzymes and genetic information.

After the last Chapter we draw some Final Remarks and we interview a Brazilian environmental researcher. Various options for Further Reading are offered, including Web Sites for Self Learning. We also provide five Appendices which complement the information in the text.

CHAPTER 2

The Electronic Structure of Atoms

2.1. Preliminary Considerations

Our present notion of an atom is that of a cloud of electrons (negatively charged) surrounding a small nucleus of positive charge. The nucleus may be thought of as a sphere of approximately 1.6×10^{-15} m diameter, or 1.6 fm, to 15 fm for the heavier atoms. More details on the constitution of the atomic nucleus shall be given in Chapter 8, which deals with this matter.

Most of the negative charge (approximately 90%) may be found within a sphere of some 0.25 to 1.25×10^{-10} m radius from the nucleus. No doubt this volume depends on the atom; the hydrogen atom has a "volume" well smaller than uranium, for instance. Thus, it is arbitrarily given the denomination of volume of the atom a sphere of 0.25 to 1.25×10^{-10} m radius. The remaining 10% of the negative charge will decay asymptotically away from the nucleus. If we think, for comparison purposes, of the nucleui and atoms as spheres, the nucleus occupies a volume some 100,000 times smaller than the atom itself.

What we call *positive* or *negative* charge in this context is arbitrary, but it is true that equal charges repel each other and opposite charges attract each other. In fact, *charge* itself is a property of elementary particles which we detect solely by the *interaction* between charged pieces of matter.[1]

2.2. The Bohr Model for Atomic Structure

Other than concepts, we need in exact sciences models which through mathematical relations may reproduce what we observe and measure in nature. For atoms, Bohr (Niels Henrick David Bohr, Denmark, 1855–1962) introduced such a model in 1913, as we describe below.

Human beings' minds feel comfortable when constructing new concepts by analogy with familiar previous notions. Thus, in the early years of the twentieth century, to imagine atoms as small planetary systems was acceptable. What was puzzling, at the same time, was that particles of opposite charge (nuclei and electrons) would not collapse by mutual electric attraction, nor escape from each other due to the centrifugal force. (But the planets and the sun do not collapse either by gravitational attraction, and neither do the planets escape).

The distances and masses involved are certainly different: for instance, the mass of our sun is 1.99×10^{30} kg while the mass of the hydrogen atom nucleus is 1.67×10^{-27} kg; thus, for atomic systems, gravity has a very minor importance. Incidentally, the mass of the universe is presently estimated at 3×10^{52} kg.

In Bohr's model the electrons of an atom describe circular orbits around the nucleus. As early as 1900, Planck (Max Karl Ernst Ludwig Planck, Germany 1858–1947) had introduced the notion that for microscopic systems the energy could have only certain definite values, meaning that not all values were possible. Thus, for such cases energy was *discrete,* rather than *continuous.* This is the concept of *quantisation* of energy. In such a way, in Bohr's model only some orbits were possible, commanded by the so-called *principal quantum number, n.* The number n may have one of the integer values 1, 2, etc. The smaller the value of n, the closer the electronic orbit to the nucleus. For instance, the first orbit radius for the hydrogen atom (for $n = 1$) has the value 5.29167×10^{-11} m. This value is given the name a_0. The value of a_0 is a universal constant called *Bohr radius* and appears in several formulae in atomic structure theory.

In the case of the hydrogen atom, the electronic energies depend solely on the number n, through the formula:

$$E_n = - (1/2n^2) \text{ a.u.} \tag{2.1}$$

Equation (2.1) is expressed in atomic units (a.u.). In the MKS unit system 1 a.u. = 4.36×10^{-18} J (or 27.21 eV). Some textbooks will show the formula $E_n = -(1/n^2)$ R, so that R = 2a.u. The number R is called the Rydberg constant. Other units may also be chosen for equation (2.1).

Formula (2.1), published by the Danish physicist Niels Bohr in 1913, remains valid today. The lowest possible energy value for the electronic energy of the hydrogen atom is

$$E_1 = -0.5 \text{ a.u.}$$

As the values of n increase, the energy values E_n (called generally *energy levels*) become closer and closer (see Fig. 2.1). If n becomes very large, E_n approaches zero. Above this limit, the hydrogen atom no longer exists, but is dissociated into two free particles: a proton and an electron.

Fig. 2.1. The hydrogen atom energy levels. Energy is in eV's.

The extremely simple formula (2.1) is nonetheless restricted to a one-electron atomic system. For larger atoms as He (two electrons), Li (three electrons) and so on, the exact energy may be calculated, but much more elaborate methods are needed. There are several procedures for obtaining the *exact* electronic energy of multi-electron atoms to as high a precision as desired, but these solutions are not *analytical* as is equation (2.1). This calculation capacity, other than the efficiency of the algorithms employed, is strongly enhanced by the ever increasing capacity and speed of computers.

2.3. Atomic Functions

As described in the previous section (2.2), Bohr's formulation provided the exact values for the energy levels of the hydrogen atom. However, the circular orbits predicted at the same time are not valid in what nowadays we believe to be the atomic structure: while electrons are distributed in spherical symmetry around the nucleus, their movement is not restricted to fixed distances from the nucleus. The spherical symmetry of any isolated atom (meaning no fields neither other atoms present) is twofold obvious. Either we realise that the choice of any Cartesian axis system is arbitrary, since infinite choices are possible, or else there is no foreseeable reason for the electronic charge cloud surrounding the nucleus to prefer one region in space rather than other.

The Austrian physicist Schrödinger (Erwin Schrödinger, 1887–1961) obtained the exact solution for the hydrogen atom problem, including the spatial functions, in 1925. For the value $n = 1$ the function (often called 1s) is

$$1s = \frac{1}{\sqrt{\pi}} \exp(-r) \qquad (2.2)$$

In equation (2.2), r is the variable distance between the nucleus and the electron (atomic units are used again).

The probability of finding the electron as a function of the distance r from the nucleus, $P(r)$, for the 1s level of the hydrogen atom is shown in Fig. 2.2. The probability of finding the electron decreases

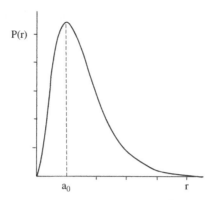

Fig. 2.2. The electronic density radial distribution, P(*r*), for the 1s orbital of the hydrogen atom.

as it approaches the nucleus. To compensate for the enormous Coulomb attraction between the two oppositely charged particles at short distances, as the electron approaches the nucleus it turns ever faster, reaching close to the velocity of light. The likelihood of finding the electron also rapidly decreases over long distances. As may be seen in Fig. 2.2, most of the time the electron will be found at a distance of a_0 (Bohr's radius) or a little more.

For atoms with many electrons and even for the hydrogen atom for levels with $n > 1$, the functions are characterised by two quantum numbers, the n number which we introduced above, and the l number, also an integer, with the name of *angular quantum number*. As in the case for the energy, the angular velocity of the electron around the nucleus may not assume any value, but only some definite values are possible. This fact is at the origin of this second quantum number.

This property was also deduced by Schrödinger in 1925. The number l may assume the values 0, 1, 2 etc., $n–1$. A very simple notation (i.e., symbol) for the atomic functions has become widely employed:

n l(as letter)

with n being the principal quantum number and l = s, p, d, f, etc. for l = 0, 1, 2, 3, etc. In such a way, the functions for each electron are labelled as:

1s,
2s, 2p,
3s, 3p, 3d,
4s, 4p, 4d, 4f,
5s,...

These symbols are used to describe the electron occupancy or *electronic configuration* of the *ground states* of atoms. The name ground state or *fundamental state* is adopted for the lowest energy state possible of the system. For example, the electronic configurations for the ground states of Li, C and Na atoms are

Li $1s^2 2s^1$

C $1s^2 2s^2 2p^2$

Na $1s^2 2s^2 2p^6 3s^1$

2.4. Excited States of Atoms

Actually, the atom, or other microscopic systems, need not be in the lowest state configuration, that is the lower energy configuration. The systems may also exist in some state with energy higher than the minimum possible. Such states are called *excited states.*

Various excited states may be generated for a given atom; these states will have energies higher than the lowest energy possible. These higher states may be produced by heating the sample or by the impact of electromagnetic radiation (for instance visible light). Possible excited states for the atoms used as examples above may be:

Li* $1s^1 2s^2$

C* $1s^2 2s^1 2p^3$

Na* $1s^2 2s^2 2p^6 3p^1$

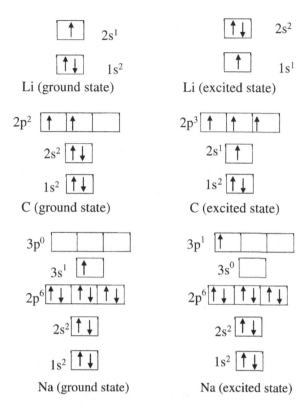

Fig. 2.3. Schematic representation of energy levels ordering and occupancy for the ground and excited states for Li, C, and Na. We employ the standard symbols of ↑ for spin up (+1/2 a.u.) and (↓) for spin down (−1/2 a.u.), respectively.

Figure 2.3 above shows schematically the occupation of the energy levels for the fundamental and excited states for the atoms Li, C, and Na. The energy level occupancy is of at most two electrons (with opposite *spins*) as mandatory by Pauli's (Wolfgang Ernst Pauli, Austria 1900–1958) *Exclusion Principle*. In Chapter 6 the concept of electron spin is discussed in more detail.

2.5. Mathematical Expressions for the Atomic Orbitals

Let us now focus on the functions which describe the behaviour of each electron in an atom. The common name for these functions is *atomic orbitals.* A rather simple mathematical form for the atomic orbitals was suggested by the American scientist J.C. Slater (John Clarke Slater 1900–1976) in 1930:

$$f_S = N_S\, r^n \exp(-ar). \tag{2.3}$$

In the former exponential expression for the atomic orbital f_S, the exponents n and a are peculiar for each atomic species; N_S is a number depending on n, a, and universal constants. These functions were employed for years and are still widely applied in the study of the electronic structure of atoms.

It is important to stress that the numerical values for n, a and N_S are specific for each orbital and different for every atom; even more, the values may be different even for excited states or ions of the same atom.

The connection between the former explicit functions f_S and the atomic functions which we name above (Section 2.3) is as follows:

i) The exponent n in equation (2.3) is the very same principal quantum number n.

ii) The angular quantum number l (also known as azimutal or orbital quantum number) is incorporated by multiplying f_S by a function depending on the angular variables, θ and φ. The set of variables r, θ and φ are the *spherical coordinates*, appropriate for the description of the movement of electrons around the nucleus. The connection between these coordinates and the more common Cartesian coordinates is shown in Appendix I. The angular functions, Y_{lm}, called *spherical harmonics,* depend on the quantum numbers l and m (see below). Thus we have the complete atomic orbital ψ_{nlm} as

$$\psi_{nlm} = f_S(r)\, Y_{lm}(\theta,\varphi) \tag{2.4}$$

iii) The number m which appears in equation (2.4) is called the *magnetic quantum number.* As the name suggests, the number m comes in experimental evidence solely when the atom is submitted to a magnetic field.

Table 2.1. Illustration of two alternative notations for atomic orbitals.[a]

ψ nlm			nl		
Ψ_{100}			$1s$		
Ψ_{200}	Ψ_{210}		$2s$	$2p$	
Ψ_{300}	Ψ_{310}	Ψ_{320}	$3s$	$3p$	$3d$

[a] It is presumed that the atom is not submitted to a magnetic field so that $m = 0$ throughout.

The correspondence between the Ψ_{nlm} and the notation nl is shown in Table 2.1.

Later, in 1950, the alternative use of Gaussian (Gauss, German mathematician, 1777–1855) type functions, f_G, was introduced by S.F. Boys and G.G. Hall. These functions have some advantages over the Slater type functions, f_S. The Gaussian type functions are easier to handle mathematically and can, eventually, with moderate increase in computational time, provide results as accurate as those obtained with the Slater type set.

The explicit form for a Gaussian type function is

$$f_G = N_G\, r^{n'} \exp(-a'r) \qquad (2.5)$$

Similar to the Slater type orbitals, n' and a' are appropriate exponents for each orbital for a given atom and may even change for an excited state or an ion of the same atom. Again, N_G is a constant factor.

The Russian chemist Mendeleev (Dmitri Ivanovich Mendeleev, 1834–1907) presented a rational classification for the structure and properties of the elements as early as 1869. The periodic table (see Appendix II) of the elements constructed by Mendeleev is a beautiful example of symmetry and order in nature.

2.6. Neutral Atoms

Let us call the (positive) nuclear charge, Z. The value of Z is equal to the number of protons in the nucleus, each having the value of +1 a. u.

Thus Z is the total charge of a particular nucleus, say $Z = 1$ for H, $Z = 2$ for He, $Z = 3$ for Li, $Z = 20$ for Ca, $Z = 92$ for U and so on. It is common to write the number of protons of an atom (also called the *atomic number*) together with the symbol (see Chapter 8). The reader is invited to examine a periodic table which shows the one hundred or so *elements* (different atomic species) of which the world we live in is made.

The periodic table shows all the elements we know and even some we do not know (yet?). We draw attention to some very important families, such as, for instance, the alkaline elements, starting with Li (very reactive and unstable as free atoms); the halogens, starting with F, some of them being gases and present everywhere in nature; and the noble gases, initiating with He. This last series was given the name of *noble* because of their reluctance to react with other elements; nowadays, they are not that "noble" since all kinds of compounds have been prepared by modern chemistry with them.

2.7. Charged Atoms or Ions

For an atom, Z equals the number of electrons, with a charge of -1 a. u. each, so the atom is *neutral*, that is, its charge is 0 a. u.

Still, stable atomic species need not be neutral. They may be positively charged (there are less electrons than the value of the nuclear charge Z) and we call them *positive ions* or *cations*. Alternatively, the number of electrons may be more than the value of Z and then we have a *negative ion* or *anion*.

As a matter of fact, nature often prefers ions as the most stable species as compared to the neutral atoms. So the Na^+ ion is stable as a part, for instance, of the common salt, NaCl, either solid or in solution, while the neutral Na atom explodes and burns rapidly when exposed to air (it may be kept as a free metal protected from contact with oxygen, water, etc).

Similarly, chlorine, Cl, is stable as the Cl^- ion in NaCl or as the molecule Cl_2, while free chlorine is very reactive and is employed for this reason in sanitary water.

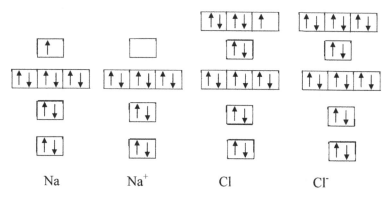

Na   Na$^+$ Cl   Cl$^-$

Fig. 2.4. Energy levels and electron distribution for the atoms Na and Cl and the corresponding Na$^+$ and Cl$^-$ ions.

Below we write the neutral atoms Na and Cl and the corresponding ions Na$^+$ and Cl$^-$:

$$\text{Na } 1s^2\,2s^2\,2p^6\,3s^1 \qquad\qquad \text{Na}^+\,1s^2\,2s^2\,2p^6$$

The former notation for the Na$^+$ cation implies that for the formation of the positive ion, an electron was removed from the higher occupied orbital 3s.

For the case of chlorine we have:

$$\text{Cl } 1s^2\,2s^2\,2p^6\,3s^2\,3p^5 \qquad\qquad \text{Cl}^-\,1s^2\,2s^2\,2p^6\,3s^2\,3p^6$$

The chlorine anion symbolised above is easy to form, since the extra electron comes to occupy the lowest orbital, still available, *3p*.

In Fig. 2.4. we show the electron distribution for Na and Cl and the corresponding ions.

In the next chapter on molecular structure and throughout the book we shall return to deal with the role of electrons.

CHAPTER 3

The Electronic Structure of Molecules

3.1. Preliminary Considerations

The word 'molecule' derives from the Latin word *moles*, meaning small mass; the present use of the word seems to have been introduced by the French philosopher, scientist and mathematician René Descartes (1596–1650).

We may extrapolate our notion of an atom, as discussed in the previous chapter, imagining a molecule as a set of nuclei imbibed in a cloud of electrons. But what holds the positive nuclei together in a stable arrangement? The answer is, the negative charge of the electrons. The manner in which nuclei and electrons are distributed in space to form a molecule and to maintain its stability is a vast field in the study of molecular structure.

An estimate of the number of conceivable compounds is over 10^{60} and every year some 10^6 new molecules are added to the list. These are either synthesised by man or isolated from nature, including living beings. Chemical science has gone a long way in rationalising, classifying and naming the structure of such an enormous variety of substances. In the following we present some selected topics in this matter.

3.2. The Chemical Bond

In nature, systems tend towards stability. Atoms form bonds, in search of increasing stability. Noble gases are less prone to form bonds since they have the so called *valence shell* complete, thus they tend to remain as isolated atoms.

The understanding of the chemical bond has gained much insight for many years through the *octet rule*, proposed by Gilbert Newton Lewis (USA, 1875–1946). According to this rule, when the outer shell of the atom (the valence shell) has eight electrons, the atom has a stable electronic configuration. Exceptions to this rule are the H^- ion and the He atom, both stabilised with two electrons in the last shell. Hydrogen is the most abundant element in the universe and is associated with the formation of stars. Helium (He) is a mono-atomic gas, colourless and odourless, which is much employed for filling balloons.

Noble gases, possessing eight electrons in the last shell, are certainly reluctant to react with other elements. These gases are components of the earth's atmosphere in small proportions. The first to be discovered was argon (Ar). Argon means lazy or inactive in ancient Greek. It gained this name because many attempts to make it react with other elements were unsuccessful. Finally, in 1962, Neil Bartlett (England and USA, 1932–2008) was able to prepare the first noble gas compound for xenon (Xe): $Xe[PtF_6]$.

Atoms get together attempting to complete the valence shell, sharing electrons. As they approach each other it is as if they were considering the advantages or disadvantages of remaining together. The atoms will form a bond if the total energy of the new structure is lower than the energy of the separated atoms.[1] A variety of possible structures may be formed in such a way.

The octet rule has been an auxiliary to chemists for a long time and still provides a useful rationale to explain structures. Still, the energy criterion provides a more general and powerful analysis of chemical bond formation.

In the following section we illustrate the formation of a chemical bond with the example of diatomic molecules, which is the simplest case.

Chemical bonds are links between the atoms, which form molecules, basic structures of substances or compounds. In nature there are just over 100 chemical elements. The combination of these elements creates an enormous variety of substances. The linkages formed are not merely random arrangements, but obey energy and affinity criteria.

3.3. Diatomic Molecules

Diatomic molecules are the simplest possible, but are by no means the least important, since many of them play crucial roles in our world. We breathe O_2. The main gas in the earth's atmosphere is N_2. The hydrogen molecule, H_2, is the simplest neutral molecule (H_2^+ is the simplest molecular ion). Both H_2^+ and H_2 were the molecular systems object of the first quantum calculations (see Chapter 6).

Hydrogen is also a source of clean energy, as described in Chapter 10. One of the features that the former examples of diatomic molecules have in common is that both atoms in the molecule are identical, being thus called *homonuclear* diatomic molecules.

But, indeed, the two nuclei forming a diatomic molecule may be different, hence we also have *heteronuclear* diatomic molecules. Some relevant examples are: NaCl (sodium chloride, or common salt), CO (carbon monoxide, a toxic gas present in the atmosphere due to incomplete combustion in the exhaust gases of automobiles).

As discussed in Chapter 2, the electrons in atoms are accommodated in atomic orbitals. When the valence shells of two atoms approach each other, an interaction or interference between the two electron clouds occurs. The outcome of this interference may be constructive, leading to a *bonding molecular orbital* (see Fig. 3.1) or destructive, that is, an *anti-bonding molecular orbital* is generated (Fig. 3.2).

When the atoms are separated (usually one says that they are at infinite distance) they behave as isolated atoms (Fig. 3.3). As the inter-nuclear distance, R, decreases they begin to 'feel' the presence of each other through the so-called van der Waals forces (Johannes Diderik van der Waals, The Netherlands, 1837–1923). These forces may be attractive or repulsive and have a strength proportional to R^{-6}.

The two atomic, 1s, orbitals interfere, forming two molecular orbitals, one bonding, symbol σ in Fig. 3.1 and one anti-bonding, Fig. 3.2, represented by σ^*. The electrons are accommodated in the bonding molecular orbital and gain more space for their movement.

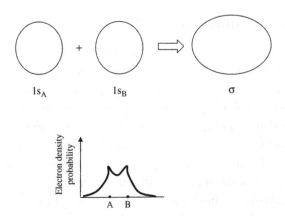

Fig. 3.1. The figure shows two 1s atomic orbitals forming a bonding orbital (denominated as σ). Below we show the electron density distribution for the H_2 molecule formed.

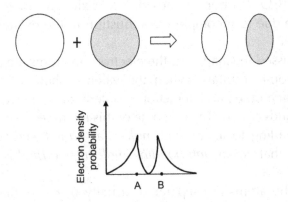

Fig. 3.2. A similar sketch as in Fig. 3.1, but in this case the molecular orbital formed is anti-bonding and will not hold the molecule together. The anti-bonding orbital is denominated σ^*.

This leads the kinetic energy to decrease until it is at a minimum, while the potential energy reaches a maximum.

When R decreases further, there is less room for the electrons to move, thus they move faster and the kinetic energy rises, while the potential energy decreases. The van der Waals forces cease to be predominant and short-range, more abrupt forces take control.

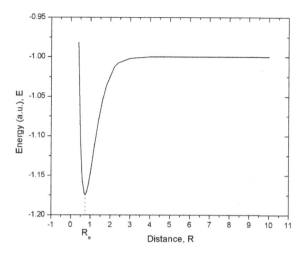

Fig. 3.3. The evolution of the total energy, E, as two H atoms approach along the internuclear distance, R. The equilibrium distance between the nuclei is denominated R_e.

Potential energy becomes ever more negative. The electron-electron repulsion is overcome by the electron-nucleus attraction. The electron density accumulates between the nuclei, allowing the atoms to approximate. At some point, the nucleus-nucleus repulsion prevents further approximations between the atoms. Then the attractive and repulsive forces reach equilibrium and the chemical bond is established. The distance between the atoms is called the equilibrium distance. The difference between the total energy at infinite distance and the total energy at equilibrium distance is denominated equilibrium dissociation energy.

A further consequence of the chemical linkage is that the electrons are confined in a reduced volume, which induces them to have an increased kinetic energy. We are here appealing to *Heisenberg's uncertainty principle*, one of the building blocks of quantum mechanics (see Chapter 6). Indeed, since the electrons are better localised, their velocity increases.

Figure 3.4 attempts to explain in some detail the event of two atoms approaching each other from a long distance, to which we give the symbol R, to the point of main interaction and then, even

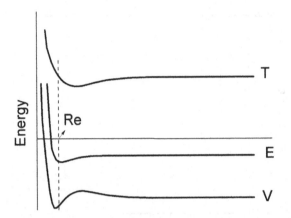

Fig. 3.4.   Two approaching and interacting H atoms along the inter-atomic distance R. E is the total energy, T is the kinetic energy and V is the potential energy. R_e is the equilibrium distance for the H_2 molecule.

closer, the two atoms almost colliding. In this simple example, the discussion refers to the formation of the H_2 molecule from two hydrogen atoms. The figure shows the evolution of the total energy, E, of the system as the value of the inter-atomic distance, R, varies.

If somehow we forced the atoms to get closer than the equilibrium distance, R_e, the potential energy, V, will have a sharp increase due to the repulsion between the nuclei. This will force the electrons to move ever faster, increasing their kinetic energy, T, and the whole system becomes unstable. The total energy, E, grows and becomes positive.

Bonding between atoms is the basis of all chemistry; therefore, the need to understand the process of formation of bonds. We should not overemphasise the role of electrons in the bonding. The nuclei have a crucial importance as well. As we shall discuss in Chapter 8, it is the nucleus than determines the behavior of atoms. The study of iso-electronic systems (atoms or ions with the same number of electrons) shows that the increase of the nuclear charge diminishes the total energy and raises the electron velocities, with important consequences for the formation of chemical linkages.

In the following sections we shall give attention to the various types of chemical bonds.

3.4. The Metallic Bond

The metallic bond arises when many atoms of a given metal (see also Chapter 5, Secs. 5.2 and 5.3) lose electrons at the same time. The electron 'sea' generated in this manner stabilises the cations so created. The interaction between the free electrons and the cationic skeleton determines the crystalline structure of metals and a series of characteristic properties. These metallic properties are:

- Characteristic metallic shine
- Resistance to traction
- High electric and thermal conductivity
- High density
- Malleability (they can take the form of sheets)
- Ductility (they can form wires)
- High boiling and melting temperatures.

Metals can mix, forming alloys with other metals or other elements. For instance, 18 carat gold is an alloy with 75% gold (Au) and 25% copper (Cu). Bronze is an alloy of copper and tin (Sn).

Steel is a combination of iron (Fe) and carbon, with the proportion of carbon varying up to 2%. With a low proportion of carbon, the steel is easier to manipulate into diverse forms. If the proportion of carbon is increased, the resistance is major and train wheels, for instance, may be produced. With 2% carbon, tools such as knives may be made.

Other examples of alloys are: stainless steel (iron with chromium, Cr), the old dental amalgam (Hg, Ag and Sn) and brass (Cu and Zn).

3.5. The Ionic Bond

The ionic bond occurs through electrostatic attraction between two ions of opposite charge. For the bond to be formed, a low electronegativity element, such as a metal, donates an electron to a high electronegativity atom, generally a non-metal (frequently, a halogen). Both elements acquire a charge (become ions) and strongly attract each other through electrostatic interaction.

The octet rule allows a simple explanation for such ionic bond formation. Let us use sodium and chlorine as examples. The electron shells for both neutral atoms are:

$$_{11}Na \; K = 2; \quad L = 8; \quad M = 1$$
$$_{17}Cl \; K = 2; \quad L = 8; \quad M = 7$$

The transference of an electron from Na to Cl produces the stable configurations:

$$Na^+ \; K = 2; \quad L = 8$$
$$Cl^- \; K = 2; \quad L = 8; \quad M = 8$$

Thus a skeleton of Na^+ and Cl^- ions forms the crystal salt NaCl and the same ions are present when the salt is dissolved in water.

Let us list just a few other examples of ionic compounds: sodium carbonate, Na_2CO_3, sodium bicarbonate, $Na(HCO_3)$, copper sulfate Cu_2SO_4, iron dichloride $FeCl_2$, iron trichloride $FeCl_3$, calcium carbonate $CaCO_3$, potassium nitrate KNO_3, lithium nitrite $LiNO_2$, etc. Some properties of ionic compounds are:

- Electric conductors when melted or dissolved in water
- High melting and boiling points
- Soluble in water
- Solids at room temperature
- Form well-defined crystalline networks
- Hard but break easily on impact
- Form planar surfaces when broken.

3.6. Covalent Bonds

In the case of the *covalent bond* the atoms *share* electrons in search of completing the valence shell, contrarily to the case of the ionic bond, in which electrons are *transferred* from an atom to another. This sharing of a pair of electrons need not be strictly half/half; depending on the electronegativity of the atoms involved, the bond may be somewhat polarised in the direction of one of the atoms.

For instance, oxygen has six electrons in the valence shell and needs two electrons to complete the octet. The hydrogen atom needs one electron to complete the K shell. In consequence, one oxygen atom shares two electrons with two H atoms, forming the vital water molecule.

The water molecule can be represented through the Lewis diagram as in Fig. 3.5.

The Lewis description does not give information on the geometry of the molecule; it merely shows the distribution of the valence electrons.

The most common type of covalent bond is the *single* bond, in which one pair of electrons is shared by two individual atoms. The examples are extremely numerous. For instance, in methane, all the four C–H bonds are single bonds; in ammonia, NH_3 there are three single N–H bonds and in the gas F_2 there is one single F–F bond.

The sharing of a pair of electrons is called a double bond. For instance in carbon dioxide, CO_2 we find two double bonds, i.e., O=C=O. We refer to this gas of crucial environmental importance in Chapter 10, section 10.5. In the oxygen gas, O_2 we have also a double bond, O=O.

Triple bonds are also common. Hydrogen cyanide, HCN, has a triple bond between carbon and nitrogen, C≡N. Cyanides are among the most lethal poisons known to man. The abundant nitrogen gas, N_2, prevailing in air, also has a triple bond, N≡N.

The covalent bond is in fact the *standard* type of bond, which is responsible for the formation of common molecules. Very large molecules may be formed through this type of linkage. A very good

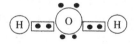

Fig. 3.5. Lewis diagram for water; electrons are represented by dots and bonding linkages by the boxes.

example is the *protein* α-keratin, the basic constituent, 70% of the composition of *hair* (in Chapter 11 we provide more information on the chemistry of proteins). A single hair filament is formed by thousands of α-keratin chains, arranged in spiral form. The interaction between the proteins gives the different nuances of hair form: straight, wavy, and curly. Both the form and colour of hair has a *genetic* background, not forgetting the effects of aging, of course. Nowadays, cosmetic chemists and hairdressers do not seem very concerned either with genetics nor with age.

The characteristics of molecular covalent compounds are:

- They may exist as solids, liquids or gases
- Melting and boiling points lower than for ionic compounds
- Normally not electric conductors.

At this point let us stress the main differences between ionic and covalent bonds. In an ionic bond atoms are kept together by the electrostatic attraction between ions with opposite charges, while in a covalent bond the link is maintained by sharing electrons.

3.7. Hybridisation

Hybridisation of atomic orbitals is a simple mathematical method to better describe the bonding in some molecules. It consists of the mixing of atomic orbitals with similar energies, leading to new orbitals with energies intermediate to the original ones. This technique has efficiently satisfied the interpretation of molecular geometries encountered experimentally. In what follows we shall employ the notion of the *spin* of the electron; each orbital may contain, at most, two electrons, one with spin 'up' (↑) and the other with spin 'down' (↓). The notion of the spin of the electron and its incorporation into the theory will be explained further in Chapter 6.

In some atoms the orbitals of the sub-shells *s* and *p* mix, forming the hybrid orbitals *sp*, *sp²* and *sp³*. These new orbitals satisfy the known geometry of the molecules. Below we explain and give examples for each type of hybridisation.

3.7.1. Hybridisation sp³

This case is clearly explained by the carbon atom in methane, CH_4, which has tetrahedral geometry, with the carbon atom at the centre of the tetrahedron and each H atom in one of the vertices. The ground state of the carbon atom has the electronic configuration $1s^2 2s^2 2p_x^1 2p_y^1$, schematically:

$$C = \frac{\uparrow\downarrow}{1s}\frac{\uparrow\downarrow}{2s}\frac{\uparrow}{2p_x}\frac{\uparrow}{2p_y}\frac{}{2p_z}$$

Having only two unpaired electrons, carbon would be apt to form just two single bonds. The first step to explaining the behaviour of the carbon atom is to promote one electron from the 2s orbital to the available $2p_z$ orbital. This gives lieu to the *excited* carbon atom:

$$C^* = \frac{\uparrow\downarrow}{1s}\frac{\uparrow}{2s}\frac{\uparrow}{2p_x}\frac{\uparrow}{2p_y}\frac{\uparrow}{2p_z}$$

If we assign the same intermediate energy to the four single occupied orbitals, we arrive at the electronic distribution which accounts for the four equal bonds of C in methane. Thus, we have:

$$C^* = \frac{\uparrow\downarrow}{1s}\frac{\uparrow}{sp^3}\frac{\uparrow}{sp^3}\frac{\uparrow}{sp^3}\frac{\uparrow}{sp^3}$$

Methane is indeed tetrahedral. Figure 3.6(a) emphasises the sp^3 hybridisation. Methane is a gas present in the earth's atmosphere and it absorbs heat even more than carbon dioxide. It is emitted by the anaerobic metabolism of cattle and the putrefaction of organic materials.

3.7.2. Hybridisation sp²

The sp^2 hybridisation occurs when one of the p orbitals is left out of the hybridisation. A clear example is the molecule of ethene ($H_2C=CH_2$),

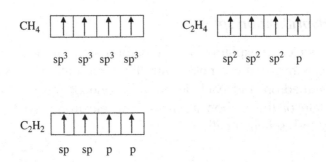

Fig. 3.6. Three cases of hybridisation: (a) methane; (b) ethene, and (c) acetylene.

which has a double bond between the carbon atoms. The electron distribution for the carbon atom is then

$$C^* = \frac{\uparrow\downarrow}{1s} \frac{\uparrow}{sp^2} \frac{\uparrow}{sp^2} \frac{\uparrow}{sp^2} \frac{\uparrow}{p}$$

The three hybrid sp^2 atomic orbitals form three σ bonds with angles of 120° between them, two pointing to the hydrogen atoms and the third to the other carbon atom. The remaining non-hybridised p orbital forms the second (π) bond linking the carbon atoms. Figure 3.6(b) shows the distribution of the atomic orbitals in the ethene molecule.

The ethene or ethylene molecule is an intermediary in petro-chemical industry for the production of polyethylene and other plastic materials.

3.7.3. Hybridisation sp

This is also a common electron distribution in some molecules. Only one of the p orbitals is combined with the s atomic orbitals, the other two p-type orbitals remaining available to form two π bonds. Many molecules have triple bonds and the most familiar case is acetylene (HC≡CH) as illustrated in Fig. 3.6(c). This gas has been used for many years in the acetylene gas lamps.

3.8. The Aromatic Bond

It was for the very common benzene molecule that the concept of the aromatic bond originated. Benzene is a liquid at room temperature, inflammable, colourless and has a sweet and agreeable aroma. But the vapours are toxic, and may even lead to loss of consciousness. Long exposure may produce leukopenia and cancer.

Benzene is widely employed in industry as a solvent or as raw material for products such as paints, plastics and gasoline.

Benzene is the most common among the aromatic compounds. In Fig. 3.7 we show its planar geometry, all carbon atoms having sp² hybridisation so the C–C–C angle has 120°. This forms the σ skeleton of the ring. In addition, we have the remaining p orbitals, one on each C atom, which will generate the π framework.

The single and double bonds are alternating; we say these are conjugated double bonds. In fact, there is no way to assign either double or single bond character to any particular pair of carbon atoms. Thus the inter-atomic CC distance is intermediate between the double and the single bond. This is represented by a circle in Fig. 3.7.

A classical notion to describe the fact that neither single nor double bonds can be ascribed to a particular CC link in benzene was denominated *resonance.* This would give the idea that the molecule was jumping from one configuration to another. In fact the molecule is neither of them but an intermediary permanent structure.

Many other hydrocarbon rings have an aromatic character, such as naphthalene, anthracene, etc. The property of cycles with neither

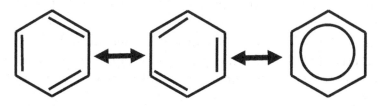

Fig. 3.7. Illustration of the resonant (or aromatic) character of the benzene CC bond.

a single nor a double bond has received the name of *aromaticity*, since benzene is aromatic.

3.9. Intermolecular Forces

Molecules also interact with other molecules, this interaction having the denomination of *intermolecular forces*. Normally, these forces are weaker than the atom-atom interactions inside a molecule (often called *intramolecular forces*), but we need to consider them when understanding the macroscopic properties of a substance.

Intermolecular forces are responsible for the different physical states of matter; otherwise all substances would be gaseous. These forces arise through the electrostatic attraction between electron clouds and atomic nuclei.

3.9.1. *Dipole moment*

To start our discussion, let us consider the polarity of molecules. The electronic cloud surrounding a molecule, say a diatomic molecule, need not be symmetrical. Depending on the electronegativity of each atom, it will attract more or less electrons. Electronegativity describes the ability of an atom to attract electrons towards it. For instance, in the hydrogen fluoride molecule (HF), fluorine, being more electronegative than hydrogen, the pair of electrons that both atoms share shall be closer to fluorine most of the time. Thus, a partial negative charge will exist in fluorine and hydrogen will consequently have a partial positive charge. The bond between H and F is a *polar* covalent bond. As a consequence, the HF molecule has a *dipole moment*, with the electronic cloud more centered in the region of the fluorine atom.

The value of the dipole moment (usual symbol μ) can be measured or calculated and constitutes a property characteristic of each molecule. For instance, the value of μ for HF is of 1.94 debyes (D).

On the other hand, if we consider the symmetric H_2 molecule, there is no physical basis for the electrons to prefer either atom. In other

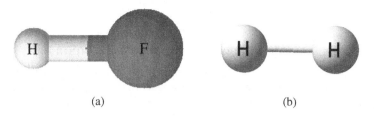

(a)   (b)

Fig. 3.8. Representations of the electronic cloud distribution around (a) the HF molecule and (b) the H_2 molecule.

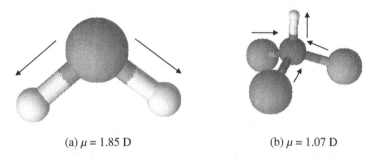

(a) $\mu = 1.85$ D   (b) $\mu = 1.07$ D

Fig. 3.9. The dipole moment vectors and μ (D) vales for two molecules: (a) water; and (b) chloroform.

words, the dipole moment of the hydrogen molecule is $\mu = 0$ D. Figure 3.8 illustrates what we have described for the electronic distribution for the HF and H_2 molecules.

One may also represent the dipole moment as a vector. The vector is defined as pointing from the negative pole towards the positive pole. Polyatomic molecules may certainly have dipole moment as well. The vector representing the total dipole in these cases is the resultant of the composition of the moment vectors of each bond. According to the presence or absence of dipole moment, molecules are classified as *polar* or *non-polar*. In general, molecules of high symmetry are non-polar, benzene being an example (Fig. 3.7). Figure 3.9 shows the dipole moment vectors and μ values of two very common polar molecules.

We have commented so far on molecules which by their composition and form possess a *permanent* dipole moment. But a charged species (e.g., an ion) may produce a distortion of the electronic cloud of a molecule otherwise non-polar. In the sketch below we show a sodium ion inducing a charge deformation in a nitrogen molecule:

$$Na^+ \quad N^{\delta-} \equiv N^{\delta+}$$

In this case we say that the N_2 molecule has an *induced* dipole moment. We add that a polar molecule may also induce a dipole moment to a non-polar species.

In what follows we describe the principal intermolecular interactions.

3.9.2. *Hydrogen bonds*

This is a very particular type of interaction between polar molecules, which is quite strong indeed, sometimes more than 10% of the value of common covalent intramolecular bonds.

They occur between molecules with a strongly electronegative atom linked to hydrogen atoms, such as H_2O, NH_3, and HF. A strong (hydrogen) bond is established between the hydrogen atom and the electronegative atom of another molecule, as shown in the scheme below for water:

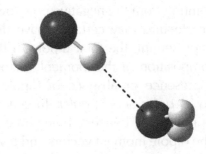

Table 3.1. Values for some selected
bond energies

Bond	Energy (kcal/mol)
H–H	432
H–O	459
C–C	346
C=C	602
C≡C	835
N–N	167
F–F	155
O···H	5 to 30

This significant O---H hydrogen bond between the water molecules explains its very high boiling point. Water is a liquid at room temperature, while the analogue molecule H_2S is gaseous in the same conditions.

The energy of a hydrogen bond is certainly lower, but comparable, to normal covalent bond energies. To facilitate an idea of the size of hydrogen bond energy, in comparison to common covalent bonds, we provide some examples in Table 3.1.

To some degree, all living processes described in Chapter 11, occur in an aqueous medium. For instance, enzyme activity is highly dependent on the water molecules present, being either free water molecules, or H-bonded to the enzyme or other molecules.

In Chap. 5, which deals with inorganic compounds, we shall give some more attention to the importance and properties of water.

3.9.3. *Van der Waals forces*

We list below the types of van der Waals forces that interact between two molecules or an atom and a molecule.

i)   *Permanent dipole* with *permanent dipole;*
ii) *Permanent dipole* with *induced dipole;*
iii) *Ion* with *permanent dipole;*

iv) *Ion* with *induced dipole;*

v) *Induced dipole* with *induced dipole;* this particular type of inter-action receives the denomination of London (Fritz Wolfgang London, Germany, USA, 1900–1954) dispersion forces; these are considered the weakest of all intermolecular forces and occur during a short time fluctuations in the charge distribution of a non-polar molecule.

CHAPTER 4

Organic Molecules

4.1. Preliminary Remarks

The distinction between organic and inorganic chemistry was the belief that organic compounds originated from living beings and could not have their origin in inorganic compounds. The latter were generated by minerals.

The synthesis of urea (an organic compound) by Wöhler in 1928 (Friedrich Wöhler, Germany, 1800–1882) by heating a water solution of ammonium cyanate (an inorganic compound), demystified the former belief. Nonetheless, the distinction between organic and inorganic chemistry persisted, although with a new definition: organic chemistry is the chemistry of carbon compounds. For some related comments, see also Chapter 5, section 5.1.

There are at a crude estimate, many thousands of organic compounds, mainly due to the capacity of carbon atoms to form linkages between themselves and other atoms, forming long chains, which may have ramifications or form cycles of various sizes.

4.2. Hydrocarbons

We shall first describe the large family of *hydrocarbons,* compounds consisting solely of two elements: carbon and hydrogen. Hydrocarbons are practically non-polar, since there is no significant difference in the polarity between carbon and hydrogen. These compounds are soluble in organic solvents (for a list of organic solvents see Table 11.1. in Chapter 11.) but not soluble in water.

According to an important feature of their structure, hydrocarbons are divided into two large groups: *Aliphatic Hydrocarbons* and *Aromatic Hydrocarbons.* Perhaps a working distinction between the two groups is that the aliphatic group contains a variety of chains but does not have benzene rings. On the other hand, the aromatic set contains one or more benzene rings. We can subsequently divide the aliphatic hydrocarbons into saturated alkanes, that contain only single bonds and non-saturated or unsaturated alkenes, which have double bonds, and alkynes, with triple bonds.

Figure 4.1. should facilitate the visualisation of the various types of hydrocarbons.

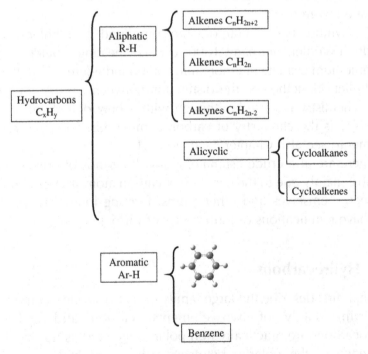

Fig. 4.1.   The classification of hydrocarbons.

4.3. Saturated Hydrocarbons: Alkanes

These may form chains with all four valences of the carbon atoms occupied by linkages with hydrogen or other carbon atoms. They obey the general formula:

$$C_nH_{2n+2} \tag{4.1}$$

The number n may take the values 1, 2, 3, and so on. When n = 1 we obtain methane, CH_4; when n = 2 the molecule is ethane, C_2H_6, and so forth.

Alkanes are relatively stable compounds, still prone to undergo chemical reactions, mainly with oxygen and halogens. The reaction with oxygen is called combustion and we shall come back to this process in Chapters 10 and 11.

The general combustion reaction is

$$C_nH_{2n+2} + (1.5n + 0.5)\, O_2 \to nCO_2 + (n + 1)\, H_2O + heat \tag{4.2}$$

Alkanes are obtained mainly from the distillation of petrol. Petrol is a non-renewable and fossil energy source. The origin is organic matter in decomposition, accumulated in deep deposits under seas or lakes. These deposits are submitted to high pressures by the earth crust and after millions of years formed the viscous dark-coloured liquid we call oil or petroleum. It is mainly employed as fuel. Other applications are as a solvent and as a raw material in the chemical industry.

An oil well produces mainly crude oil and a smaller amount of gases. There are also gas wells that generate mainly the so-called natural gases.

Depending on the proportions of different hydrocarbons in the oil, various classes of crude oil may be defined. Once the composition of a given oil is known, different fractions with specific properties may be separated. The separation process is based on the boiling point of the different hydrocarbons and is denominated *fractional distillation*. The process consists in heating the oil and passing the gases thorough long columns, so that fractions with different boiling points may eventually separate.

This process produces a fraction of light hydrocarbons to be used as gasoline in a proportion way below consumption demand, leaving large amounts of heavy oil as by-products. As a consequence, the petrol industry proceeds to the *cracking* of heavy oils portions, producing more gasoline and other light fractions with higher commercial demand.

The catalytic cracking consists of heating the petrol up to 500 °C in the presence of a catalyser. This process produces both the breaking and the rearrangements of the alkane chains, leading to shorter and branched molecules. This result is favourable for the production of good quality gasoline. Below we show an example of a cracking reaction:

$$C_{16}H_{34} \rightarrow C_8H_{18} + 4\ C_2H_4 \qquad (4.3)$$

$C_{16}H_{34}$ is an alkane found in crude oil, specifically in a fraction denominated diesel oil. Octane is a typical component of gasoline. Catalysers for the cracking are usually aluminum silicates, also named zeolites. If the heating is done in the absence of a catalyser (thermal cracking) the tendency is the preponderance of linear alkanes, not convenient to be employed as gasoline.

The best gasolines have a high proportion of branched alkanes, as for instance 2,2,4-trimethyl pentane commonly known as isooctane (see Fig. 4.2a). The isooctane molecule burns smoothly in the

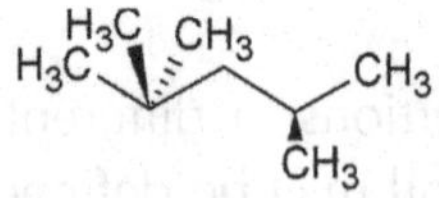

(a)

(b)

Fig. 4.2.   (a) Iso-octane and (b) n-heptane.

engine and is considered the most efficient combustible. It is used as the standard for top octane rating.

The linear alkane n-heptane (Figure 4.2b) has the worst perform-ance, with the tendency to ignite by compression decreasing the effi-ciency of the car engine. Heptane is given a value of zero in octane rating. The higher the octane rating the better is the performance of a given gasoline.

4.4. Unsaturated Hydrocarbons: Alkenes

Alkenes differ from alkanes by the existence of unsaturated bonds, or double bonds, of π character. This insaturation is at the root of signif-icantly enhanced reactivity as compared to alkanes. The general for-mula for alkenes is C_nH_{2n}, for the case of only one double bond.

Ethene or ethylene [$H_2C = CH_2$] and propene or propylene [$H_3C - CH = CH_2$] are the simplest alkenes. Ethene is employed as the initial material for the synthesis of various industrial products as ethanol, ethylene oxide, and polyethylene.

Ethene occurs in nature in fruits as a hormone. It naturally aids in the process of ripening of fruits. In fruit commerce, fruits are often transported unripe and on arrival treated with ethene.

Ethene is obtained from the cracking of crude oil and is the most produced organic chemical, raw material for various com-pounds. Perhaps most of ethylene is employed in the fabrication of polyethylene.

Alkenes tend to polymerise when the reaction is initiated with a radical (a *radical* has an unpaired number of electrons). The reaction may be initiated by breaking a peroxide $O - O$ bond and adding the radical to the ethene. The scheme below shows the initial steps for the formation of the radicals (the ancient Greek letter Δ is often employed in chemistry to represent applied heat):

$$R\text{-}O\text{-}O\text{-}R \xrightarrow{\Delta} 2\,R\text{-}O^{\bullet}$$

$$R\text{-}O^{\bullet} + CH_2{=}CH_2 \longrightarrow R\text{-}O\text{-}CH_2\text{-}^{\bullet}CH_2$$

The formed radical links to another alkene molecule, forming again a radical, leading to a chain reaction producing large chains of polyethylene. We show this process in the following scheme:

$$R\text{-}O\text{-}CH_2\text{-}^{\cdot}CH_2 \xrightarrow{CH2=CH2} R\text{-}O\text{-}CH_2\text{-} CH_2\text{-} CH_2\text{-}^{\cdot}CH_2 \xrightarrow{CH2=CH2}$$

$$R\text{-}O\text{-}CH_2\text{-} CH_2\text{-} CH_2\text{-} CH_2\text{-} CH_2\text{-}^{\cdot}CH_2 \xrightarrow{CH2=CH2} \bullet\bullet\bullet$$

Polyethylene bags are a most common feature of stores and supermarkets, although presently their use is losing terrain because of environmental pollution. Old-style bags or fancy new bags are the new vogue.

4.5. Unsaturated Hydrocarbons: Alkynes

These hydrocarbons contain one triple bond, with general formula C_nH_{2n-2}. They suffer easily addition reactions at the triple bond.

The simplest alkyne molecule is ethyne, H-C $\equiv$ C-H, more commonly denominated as acetylene. It is obtained as a side product with ethylene in the oil cracking. It is a combustible for illumination and soldering.

Acetylene is also easily produced from calcium carbide according to the reaction

$$CaC_2 + 2H_2O \rightarrow HC \equiv CH + Ca\,(OH)_2 \qquad (4.4)$$

Acetylene may explode violently when compressed, thus it is transported in cylinders under moderate pressure.

4.6. Aromatic Compounds

At the beginning of organic chemistry, the name *aromatic* was given to some fragrant substances, as benzaldehyde (almond-like odour), toluene (balsam), and benzene (sweet aroma). Soon it was observed that the chemical properties of these compounds differed from the known chemicals.

Nowadays, we employ the term 'aromatic' to denominate the family of benzene and its structural derivatives. Benzene, C_6H_6, may be represented by a ring-type conformation. In terms of atomic orbitals, one verifies that the double bonds do not admit being assigned to a given pair of carbon atoms; thus the hexagonal benzene ring is a resonant hybrid. It is this resonant character which determines the properties called *aromaticity* in modern chemistry.

Benzene is an extremely dangerous chemical for living beings. Its planar hexagonal form permits its insertion between pairs of DNA bases (DNA is described in Chapter 11) producing genetic mutations. Even short-term exposure is toxic and in laboratories should be handled with care or, whenever possible, replaced by toluene as a reagent.

Many natural chemicals posses aromatic rings in their composition (for instance morphine). A large number of synthetic compounds have benzene or similar rings as part of their structure (drugs as tranquilisers, amphetamines, etc.).

The main sources of aromatic hydrocarbons are coal and petrol. Coal is a complex structure, mainly formed by large molecules with many benzene rings connected. The heating of coal at 1000°C produces benzene, toluene, naphthalene and many other chemicals (see Fig. 4.3).

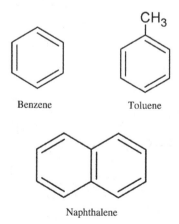

Fig. 4.3. The chemical structures of three aromatic rings.

4.7. Functional Groups

The capacity of the carbon atom to form four strong bonds with other carbon atoms, as well as with H, O, S and N allows an enormous variety of different molecules, mainly in complex living beings. A classification is facilitated defining the *Functional Groups* which occur in this vast set of different chemical species. Different molecules that contain the same functional group tend to present similar chemical properties and reactions. In Table 4.1. we present a summary of the characteristics of the different groups.

4.7.1. *Halogenated derivatives*

One or more hydrogen atoms of an aliphatic hydrocarbon may be substituted by elements from the halogen group including, fluorine, chlorine, bromine or iodine. Halogenated organic compounds are common in nature and also have several industrial applications; they are used as inhalant anaesthetics, pesticides, and in refrigeration. Notice that some of these applications are in fast decay because of atmospheric contamination (ozone depleting) or direct toxicity as is the case of the use of fluorine derivatives and chloromethane in refrigeration.

Large amounts of chloromethane are produced in the oceans from algae biomass and chlorine from the sea foam. In Fig. 4.4 we show a couple of examples of halogenated derivatives of hydrocarbons.

Dichloromethane, CH_2Cl_2 and chloroform are solvents and were used as anesthetics. Chloroform is a potential carcinogen and its use is being abandoned.

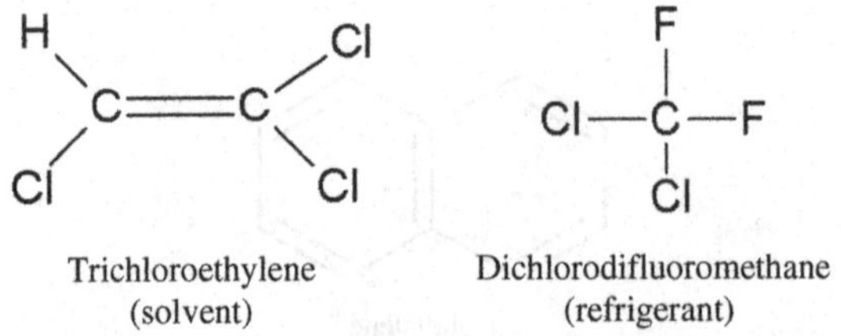

Trichloroethylene
(solvent)

Dichlorodifluoromethane
(refrigerant)

Fig. 4.4. The chemical structures of trichloroethylene and dichlorodifluoromethane.

Table 4.1. Names and characteristic chemical structures of the different functional groups.

Family	Functional group
Halogen compounds	R—X (X=Halogen)
Alcohols	R-OH
Ethers	R_1-O-R_2
Ketones	
Aldehydes	
Carboxylic acids	
Esters	
Amides	
Amines	
Amino acids	

The symbols R, R_1 and R_2 designate side chains formed by carbon atoms or other elements.

4.7.2. Alcohols

Alcohols, mainly ethylic alcohol or ethanol, CH_3CH_2OH, are present in everyday life as beverages, cleaning products, perfumes. Alcoholic beverages are controlled, restricted or forbidden, depending on the legislation and culture of different countries.

Ethanol was one of the very first organic compounds to be prepared and purified by man; it was prepared by fermentation of grains or sugar.

Methanol, CH_3OH, is employed as solvent and for the preparation of formaldehyde and is being intensively tested for the fabrication of direct methanol fuel cells.

4.7.3. Ethers

Ethers present one oxygen atom linked to two alkyl groups. There is no hydrogen bond to the oxygen, so ethers do not form acids.

Ethers have anesthetic properties, are good solvents, are raw material for syntheses and are used in perfumes. Ethyl ether, $CH_3CH_2\text{-}O\text{-}CH_2CH_3$ was employed as an anesthetic for many years.

4.7.4. Carboxylic acids

These compounds present the carboxylic functional group, –COOH; this group is linked to a carbon atom, most of the time at the extreme end of a hydrocarbon chain. The carbon atom in the COOH group has a double bond with one of the oxygen atoms, C=O, and a single bond with the hydroxyl group –OH:

$$R\text{—}C{\overset{\textstyle O}{\underset{\textstyle OH}{}}}$$

The oldest carboxylic acid known to man is acetic acid or ethanoic acid, CH_3 –COOH (see Fig. 4.5), which gives the characteristic acid taste to vinegar. Vinegar is produced by the fermentation of diverse alcoholic beverages. Any alcohol of 7% or more exposed to air will sour rapidly. The process is carried out by bacteria present in the fruits, juices and wine. To prevent this process, beverages are pasteurised or protected with antibiotics. The reaction of transformation of ethanol to acetic acid is shown in equation (4.5)

$$CH_3CH_2OH + O_2 \xrightarrow{\text{Acetobacter aceti}} CH_3CO_2H + H_2O \quad (4.5)$$

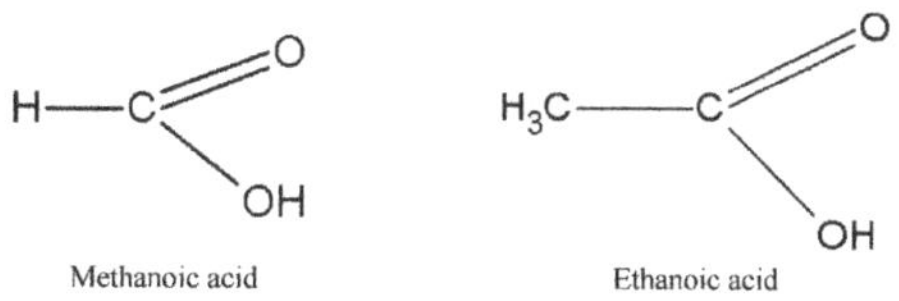

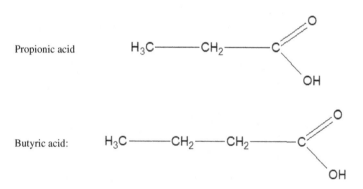

Methanoic acid   Ethanoic acid

Fig. 4.5. The structural formulas of methanoic and ethanoic acids.

The ancient Romans used to boil vinegar in pots made from lead to produce sweet liquor, not knowing that they were being poisoned.

Micro-organisms in the process of decomposition of food produce other carboxylic acids. The smell of these products, for instance cheese, is matured or rotten, depending on the case and or the point of view. These odour properties are due to acids: propionic, butyric, valeric, caproic, etc. (see Fig. 4.6).

Propionic acid H_3C————CH_2————C

Butyric acid: H_3C————CH_2————CH_2————C

Fig. 4.6. The structural formulas of propionic and butyric acids.

4.7.5. *Esters*

Esters are derivatives of carboxylic acids by the substitution of the hydrogen atom by an alkyl group, say R_2: R_1COOR_2 where R_1 is another alkyl group belonging to the carboxylic acid structure, i.e.,

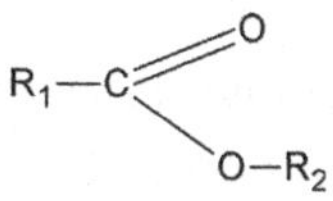

R_1—C

Esters are common in nature forming oils and fats. Esters have agreeable odor and are used in perfumery for artificial fruit essences. They are important food storage for living beings (see Chapter 11).

4.7.6. Ketones

This group of compounds has a structure, where R_1 and R_2 are alkyl groups.

$$R_1 - \overset{\overset{\displaystyle O}{\|}}{C} - R_2$$

Cyclic ketones of high molecular weight have a pleasant smell and are employed in expensive perfumes. Camphor (Fig. 4.7a) was employed for centuries in medicine and testosterone (Fig. 4.7b) is the hormone responsible for male characteristics. Cyclohexanone is the basis for the synthesis of nylon. Ordinary acetone, $CH_3 - CO - CH_3$, because of toxicity is on the way to losing its established function as a nail polish cleanser.

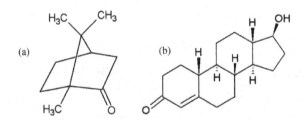

Fig. 4.7.   Chemical structures of (a) camphor and (b) testosterone.

4.7.7. Aldehydes

This class of functions presents a carbonyl group always at the end of a molecular chain, R, i.e., R −CHO. The simplest aldehyde is formaldehyde. Formaldehyde is a gas, used for the conservation and disinfection of biological tissues in an aqueous solution of 37% known as formol.

Formaldehyde polymerises with ease and the resulting polymers depend on the initial medium condition.

4.7.8. Amides

Amides are compounds derived from carboxylic acids, by the substitution of the hydroxyl group –OH for an amino group –NH_2, resulting in the functional group amide,

$$R - C \overset{O}{\underset{NH_2}{<}}$$

The molecule of urea, $CO(NH_2)_2$, may be considered as an amide from the carbonic acid, H_2CO_3. Urea is one of the final products of superior animal metabolism, excreted through urine.

The amide derived from sulfanilic acid (sulfanilamide) and other related substituted amides have relevant therapeutic applications.

4.7.9. Amines

Amines may be considered derivatives of ammonia, NH_3, by partial or total substitution of the H atoms by alkyl radicals. Several amines of natural origin, mainly found in plants, were named *alkaloids* due to their pronounced basic or *alkaline* characteristic (a basic substance tends to deliver protons, H^+). In Fig. 4.8. we show various amines.

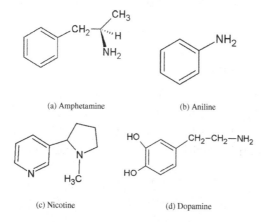

(a) Amphetamine   (b) Aniline

(c) Nicotine   (d) Dopamine

Fig. 4.8. The chemical formulae of several amines: (a) amphetamine; (b) aniline; (c) nicotine and (d) dopamine.

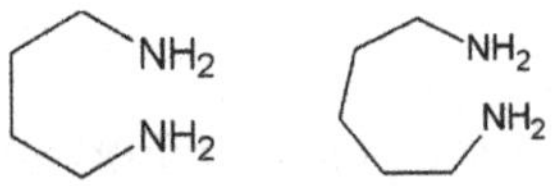

(a) Putrescine   (b) Cadaverine

Fig. 4.9.   The chemical structures of (a) putrescine and (b) cadaverine.

Some amines are formed during the decomposition of organic material (organic in the sense that they are originated from alive beings) like meat and dead fish. The odour of rotten eggs is due mainly to hydrogen sulphide, H_2S, and other organic sulphides; but the bad smell of meat and fish in decomposition is produced mainly by amines. So some of these compounds gained very appropriate names such as putrescine and cadaverine (see Fig. 4.9). Even amines of simpler composition as methylamine and diethyl amine have a smell reminiscent of rotten fish.

Some other amines have pronounced smells as well; for instance, indole has flower scents, while escatol is responsible for the bad smell of excrement. Believe it or not, escatol is used in small proportions in some fancy perfumes.

The habit of seasoning fish with lemon juice, other than that the resulting taste is agreeable for many, produces the reaction of the citric acid from the lemon, producing stable compounds without a "fishy" smell.

4.7.10. *Amino acids*

These compounds carry two functional groups: the carboxyl group and the amine group. They are the constituent units of proteins, a basic structure of living beings (see Chapter 11, section 11.2). We show an alpha-amino acid structure in the scheme below:

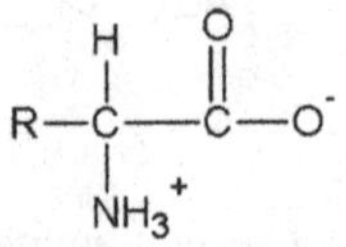

Under the action of enzymes two amino-acids bond together forming a dipeptide. These reactions may continue with the formation of polypeptides or proteins.

4.7.11. *Carbohydrates*

Carbohydrates are alkyl chains with many hydroxyl groups attached and also with functional groups as aldehydes, ketones, and carboxylic acids.

Carbohydrates are one of the main solid components of food, with a wide presence in nature. They have a variety of structures and functions as sugars (glucose, sucrose, fructose) which give a sweet flavour to food, starch, food reservoir of plants, cellulose, support structure for plants. Figure 4.10. shows some examples of carbohydrates.

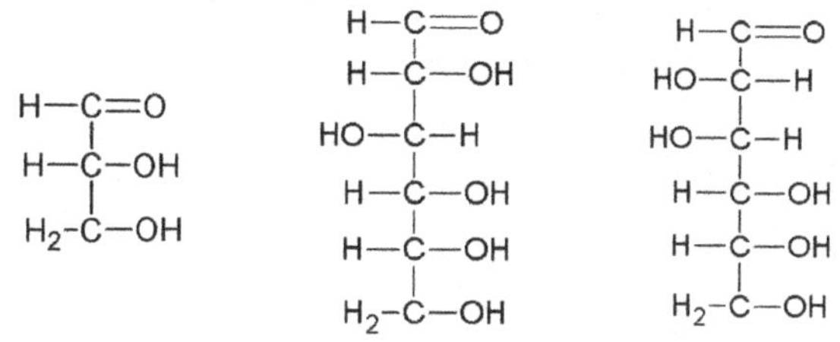

$$
\begin{array}{ccc}
 & H-C\!=\!O & H-C\!=\!O \\
 & H-C-OH & HO-C-H \\
H-C\!=\!O & HO-C-H & HO-C-H \\
H-C-OH & H-C-OH & H-C-OH \\
H_2\text{-}C-OH & H-C-OH & H-C-OH \\
 & H_2\text{-}C-OH & H_2\text{-}C-OH \\
\text{(a) Glyceraldehyde} & \text{(b) Glucose} & \text{(c) Manose}
\end{array}
$$

Fig. 4.10. The chemical structures of (a) Glyceraldehyde; (b) Glucose and c) Manose.

4.7.12. *Lipids*

Lipids are oils and fats. The difference is rather of physical consistency. At room temperature oils are usually liquids and fats solids. Vegetable oils are extracted from seeds (olive, corn, cotton, soybean, peanut, etc.). Several chemical structures belong to this family of compounds (for more information see Chapter 11, section 11.3), often a long aliphatic chain, which may be saturated or have one or several double bonds. One very frequent functional group in oils is ester.

A common industrial process is the hydrogenation of oils for the saturation of double bonds, leading to products appropriate for preparing margarine and for baking. This process improves the consistency and delays rancidity. Below we show an example of hydrogenation:

$$CH_3(CH_2)_7-CH=CH-(CH_2)_7COOH + H_2 \rightarrow CH_3(CH_2)_{16}COOH$$

oleic acid   stearic acid

Hydrogenation is nevertheless not always complete and in the process some double bonds isomerise from *cis* to *trans*. *Trans* fatty acids are rare in nature and are blamed for the rise in levels of cholesterol and triglycerides in humans, which are in turn associated with cardiovascular risk.

Our bodies do not have the capacity to produce polyunsaturated fats, so small amounts are necessary in the diet. Otherwise, saturated fats are synthesised from carbohydrates.

CHAPTER 5

Inorganic Molecules

5.1. Preliminary Considerations

There is an almost automatic association between the notion of something being *inorganic* and the elements we understand as metals: sodium (Na), copper (Cu), and so on. Gases such as nitrogen (N_2) or chlorine (Cl_2) are also considered inorganic. Nonetheless, proteins, basic constituents of living beings contain nitrogen. And sodium chloride (NaCl) is a major component of our bodies and our food. The current definition of inorganic compounds is that they do not contain the element carbon. This is what one can call a working definition; it is good as far as it is useful.

On the other hand, the word *organic*, has connotations of some biological origin or function. Actually, the separation between organic and inorganic compounds is becoming obsolete and many heterodox families of compounds are presently of interest, most of them of synthetic origin.

Let us then describe a few interesting groups of inorganic compounds, some of them very common in our daily life, other known rather to specialised chemists.

In Chapter 5, section 5.4. we shall pay close attention to water. Earth's surface and life on it is so much imbricated with this inorganic compound that we have selected it for a more detailed discussion in this book.

5.2. Metals

Several metals and alloys (homogeneous mixtures of two or more metals) have been for centuries well familiar to mankind, such as copper, iron, bronze, steel, aluminum, silver and mercury. They share some common properties. They conduct electricity and heat, have a metallic shine, dilate when heated and melt if heated further. At earth surface temperature they are solids, with the exception of mercury, which is becoming a nostalgic remembrance, as digital and liquid crystal thermometers take over.

There are other very abundant metals which do not occur as free elements in nature but combined as salts or as ions in solution. Let us mention sodium and potassium, the respective ions being Na^+ and K^+, generally with chlorine (Cl^-) as the counter-ion. These salts are present almost everywhere in living beings.

The element calcium (Ca) is a component of our bones in the form of the mineral *hidroxyapatite* or *hidroxylapatite*, with the formula $Ca_{10}(PO_4)_6(OH)_2$.

Another metal which one could say governs our lives is iron (Fe). This element plays a crucial role in our breathing function, i.e., the capture of oxygen in the lungs, the transport of oxygen by the red cells through the blood stream and the delivery to the cells to allow them the "combustion" of glucose, $C_6H_{12}O_6$, for cells' metabolic energy needs (see Chapter 10, section 10.5 for combustion of hydrocarbons). An iron ion (Fe^{2+} or Fe^{3+}) is at the centre of a molecule called *haeme* (see Fig. 5.1.), which is the active site of the haemoglobin molecule.

Oxygen binds to the Fe^{2+} ion of the haeme, which is oxidised to Fe^{3+}. When the oxygen is transferred to the cell, the Fe^{3+} ion is reduced back to Fe^{2+}.

Other than the aforementioned metals, an important number of metallic elements (often in very tiny amounts) are essential for our metabolism. These metals may be part of enzymes (see Chapter 11, section 11.3) or need to be present in the aqueous medium for given enzymes to function. There is an enormous pharmaceutical market for food supplements with vitamins, the so-called essential metals

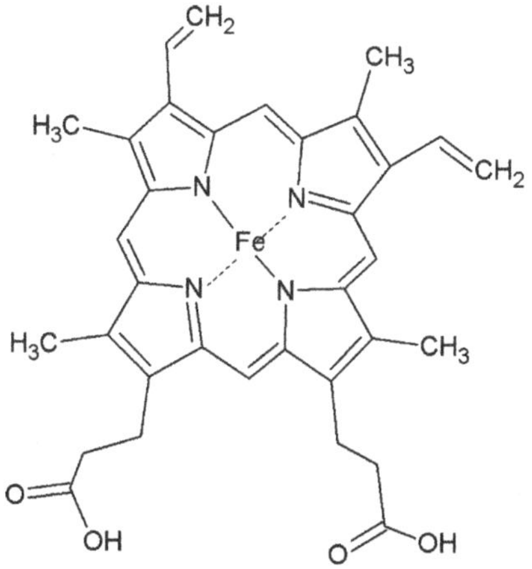

Fig. 5.1. The chemical structure of the haeme molecule, the active site of the haemoglobin protein.

and occasionally other ingredients deemed to have beneficial properties. Some of the metals found in these formulations are zinc, copper, chromium, manganese, and molybdenum (Zn, Cu, Cr, Mn, and Mo, respectively).

5.3. Electrical Conductors

Because of its abundance, high electric conductivity and stability, copper is the most widely employed electric conductor. Metals are not the only materials with conducting properties and there is an active search for non metallic electric conductors (for example, see section 5.3. below). The search for other materials has several motivations as: the reserves of copper are finite, other materials could be cheaper and lighter, and might even have conductivity higher than copper or silver.

Band theory has produced a neat and accessible explanation for the conducting properties of materials. Already in Chapter 2 we saw for atoms how the electrons are accommodated into energy levels of increasing energy, one or at most two of them in each level, in obedience to Pauli's Exclusion Principle. The bigger the molecule, the larger the number of electrons to be allocated, and the closer

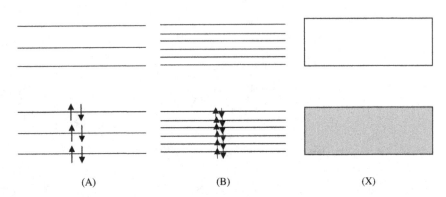

(A)   (B)   (X)

Fig. 5.2. Energy level diagrams for molecules of increasing size (thus increasing number of electrons) A, B and X. One may consider X, for all practical purposes, as having an infinite number of electrons. The lower stack of levels is occupied by electrons, while the upper stack is empty.

becomes the distance between the energy levels in the energy scale. At some stage, the distance between such energy levels may become indistinguishable. This evolution is illustrated in Fig. 5.2 for the hypothetical molecules A, B and X.

For a piece of solid metal, or other ordered materials, when the number of atoms tends to infinity (in fact in chemistry and physics this means very large), the energy levels become so close that one cannot distinguish one from another. In such a way, the electrons are embedded in a continuum. Then we speak of a band occupied by electrons. The corresponding virtual band is empty.

The top end of the occupied band is called *Fermi level* and the occupied band is known as the valence band; in this region, the electrons are bound to the nuclei. If, by some mechanism, electrons reach the empty band, the electrons become free to move through the material, the material becomes a conductor. The energy difference between the Fermi level and the conducting band is commonly denominated as *band gap.*

Further, regarding their electric conductive properties, materials are classified as insulators (there is a large band gap), semiconductors (band gap above the zero value and up to some 3 eV) and metals (the valence band and the conducting band form a continuum,

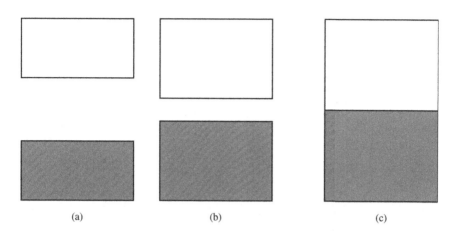

(a)   (b)   (c)

Fig. 5.3. Schematic representations of (a) insulator, (b) semiconductor, and (c) metal.

the band gap is zero). The three situations described above are illustrated in Fig. 5.3.

5.4. Sulphur Nitrides

These compounds are formed solely by diverse combinations of sulphur and nitrogen, are of strictly synthetic origin and rather unstable. We shall describe three interesting exponents of this family of chemical compounds.

S_4N_4, tetra sulphur tetra nitride, was one of the first compounds of this group to be described. It has a cage-like form as shown in Fig. 5.4a, with all four nitrogen atoms in a plane and two sulphurs above and two below that plane. The molecule is maintained stable by single S-N bonds.

The $S_3N_3^-$ anion is a planar hexagonal molecule, with the same geometry as benzene. Benzene is held together by a network of C-C single bonds plus a half bond produced by the π resonant bond over the six carbon atoms, thus the total bond value attributed to a C-C linkage in benzene is $1.0 + 0.5$, i.e., some 1.5 bond strength. In the case of the sulphur nitride chemistry, normal SN bonds are rather "weaker", a bond strength of 0.5 representing a normal SN linkage

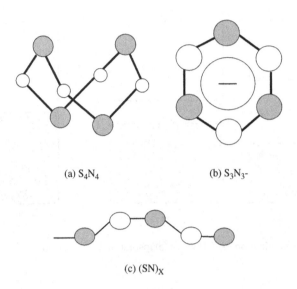

(a) S_4N_4 (b) $S_3N_3^-$

(c) $(SN)_X$

Fig. 5.4. The spatial structures of (a) the tetra sulphur tetra nitride molecule; (b) the tri sulphur tri nitride anion, and (c) a fragment of the polythiazil polymer.

(in S_4N_4 for instance). This is approximately the value (0.52) for the SN bond in $S_3N_3^-$. The contribution in this case of the π resonant system is rather modest.

A very interesting member of this family of sulphur nitride molecules is the polymer *polythiazil*, $(SN)_x$. A fragment of this polymer is depicted in Fig. 5.4(c), showing the S-N-S-N sequence in its normal planar conformation. Polythiazil is a rare example of a non-metallic *intrinsic* electric conductor; intrinsic meaning it is a conductor without the need to add other molecules (*dopants*) to stimulate conductivity. In the habitual case (poly-acetylene, for instance) conductivity is obtained through addition of other chemicals.

Among the best electric conductors are silver and copper metals. Copper cables are the main electric conductors, used everywhere in the world. The electric conductivity of copper is of 60×10^6 S/m in MKS units, the symbol S standing for Siemens. The conductivity of polythiazil is of the order of 10^5 S/m, a very high value for a non-metal non-doped material. Also, close to zero absolute temperature, $(SN)_x$ becomes a superconductor.

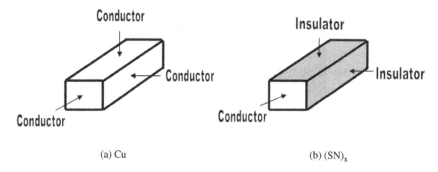

(a) Cu   (b) $(SN)_x$

Fig. 5.5. Comparison of electric conductivity and coloration of bars of copper and polythiazil.

Other than its high electrical conductivity, $(SN)_x$ is an anisotropic electric conductor. Indeed, if the polymer is prepared with the SN chains parallel to each other, its conductivity will be as noted above in the chain direction, while the transversal conductivity is of the order of 10^2 S/m, a value typical of semi-conductors. Also, while a copper bar is bright on all sides, for $(SN)_x$ we will see the metallic shine only along the chain direction. The former features have led to the denomination of materials such as polythiazil as "one-dimensional metals". Figure 5.5. compares a bar of copper and a bar of polythiazil.

Polythiazil is much lighter than copper, and would need less insulating protection due to its anisotropy. Unfortunately, it is an unstable material, but it suggests lines of research for alternative electric conductors.

5.5. Water

Since birth (as a matter of fact, before birth, in the amniotic fluid) we humans are in permanent contact-dependence with water. We indeed are dependant on other life sources as well (solar energy, oxygen, food) but in this chapter we will emphasise water. Water is an inorganic compound, H_2O, molecular weight of 18.01 g/mol, OH distances of 0.958 Å, and H-O-H angle of 104.45°. It covers more than 70% of the earth's surface, but is also present under the surface in aquifers, above the surface as vapor. For the earth's surface temperature range, while most water is in the liquid state, we can find it also as solid (ice) at the poles of our planet or the top of high mountains.

By far the oceans hold most of the earth's surface water (97%), with 2.4% in the form of ice at the polar caps, and a mere 0.6% in lakes, rivers, and creeks. The former values may have some variations at geological time scales.

The geographical distribution of water on the earth's surface is far away from homogeneity, with large surfaces, such as in the Middle East with scarce resources and other enormous countries as Brazil or the USA with plethora of water availability. Even in such fortunate cases, the geographical and annual distribution of rains is to a large extent aleatory, thus in the latter countries people also suffer from droughts and floods frequently.

Water is crucial for our survival, not only as clean drinking water, but also for agriculture, cattle growth, industry, and navigation.

The abundance and heat capacity of water protects earth from abrupt fluctuations in temperature.

Fresh water is a major need and also a controversial issue for humanity. In spite of its constant use in industry and human consumption, access to safe drinking water has improved steadily over the last decades in most of the world. This is the result of man's responsibility and technology used to recover used water for further employment. Sadly, for geopolitical reasons, a large part of the world has not benefited from this concern, either in very poor and undeveloped countries or neglected regions of technically advanced countries.

As much as 70% of available water is consumed by agriculture. In spite of the aforementioned optimistic progress, the present tendency is towards deterioration of water supplies. More and more, the lack of availability of fresh water may become a matter of conflict and competing demand. Energy sources, freshwater, food and even clean air may, at different instants, places and circumstances become major crises for humanity.

The use of such large quantities of water for agriculture should be properly understood. This water is basically recovered, through consumed vegetables, vapour elevating to clouds, or draining into rivers and aquifers. Some like to exhibit dramatic numbers for the consumption of water for each kilogram of meat. There is a problem indeed, but one should remember that cows give milk and cattle urinate.

5.6. Hydrogen Bonds in Water

Water is a very peculiar liquid, even due to the fact that it *is* a liquid. We can swim in the seas or rivers at very agreeable temperatures, never having to worry about water starting to boil (which occurs only at 100°C at sea level, usually in a pot in the kitchen). Nonetheless, the boiling point of sulphur hydride, H_2S, is as low as −60.28°C. Notice that the atomic mass of sulphur is the double of that of oxygen. Ice (ice formed form water) melts at 0°C, while solid H_2S melts at a temperature as low as −82.30°C.

We are making comparisons with sulphur hydride because the S atom is below oxygen in the periodic table. The table is in fact denominated as periodic because elements in a column tend to evolve smoothly in their properties as we descend the column (see for instance the alkaline metals, or halogens or inert elements).

On the other hand, if we move through the periodic table in a line towards the right direction, we see also a degree of gradual change from a metallic character towards increasingly non-metallic properties.

The boiling and melting points of water are abruptly dissonant of what one could expect of the place of the oxygen atom in the periodic table. Chemists have understood for many years this "anomaly": water molecules tend to associate in pairs (or in larger clusters) through the relatively strong hydrogen bonds, as we already anticipated when referring to intermolecular forces in section 3.9. In Fig. 5.6. we illustrate the stable conformation of two water molecules hydrogen bonded. The strongly electronegative oxygen atom creates an attraction with the proton of another molecule and stabilises a dimer.

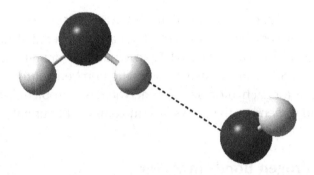

Fig. 5.6.   The $(H_2O)_2$ dimer stabilised through a hydrogen bond.

Let us only mention that oxygen is not the only atom capable of forming H-bonds. Indeed, other strongly electronegative atoms also form such kind of linkage, such as fluorine in hydrogen fluoride, HF. Oxygen and fluorine are the elements of higher electronegativity in the scale created by Linus Pauling (American, 1901–1994, awarded two Nobel Prizes). His 1935 book with Wilson can be read today with permanent and actual notions (see end of this book for Further Reading). The sulphur atom, on the other hand, has not a value of electronegativity close to oxygen and does not form hydrogen bonds in the same manner.

The "anomaly" of water we referred to before is the simple fact that water contains *de facto* molecules formed through the H-bridges with sizes even larger than simply $(H_2O)_2$ units. *Ergo*, the properties of water correspond to dimers or even larger clusters as we shall comment now.

The water molecules, through the H-bonds, tend to cluster in groups, generally of symmetrical character. Figure 5.7 shows the structures of clusters with n = 2 to n = 5 water molecules.

Clusters with a large number of water molecules may also exist, and there is evidence and calculations for them.

In liquid water these structures are extremely ephemeral. If one cluster of six water molecules is formed, it will exist for a very short time (a few femtoseconds). But the number of hexagonal structures at a given temperature is constant, thus at some other place a hexagon

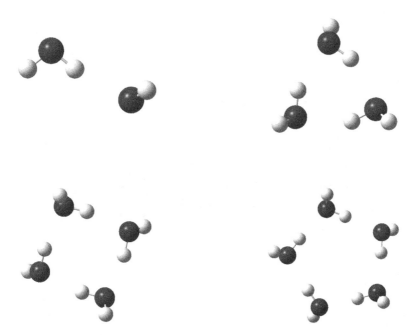

Fig. 5.7. The geometry of water clusters for n = 2 to n = 5 water molecules.

will be formed. Nonetheless, as the temperature approaches freezing point, these structures tend to stabilise, being ultimately permanent in ice. The existence of these structures in ice is the reason ice has a lower density than liquid water and thus floats.

Very promising lines of discoveries are open for future studies on the structure of water. We know now, for instance, that applied electric or magnetic fields tend to stabilise the clusters, in a similar manner as produced by the lowering of the temperature.

Another interesting feature, yet to be explained, is the preference of the clusters for specific number of water molecules participating in the more stable structures. For instance, clusters with n = 4, 8 or 12 water molecules are more stable than other clusters. This is illustrated in Fig. 5.8.

Indeed, one can observe in Fig. 5.8 that the calculated energy values for the clusters with the number of water molecules 4, 8, and 12 are somewhat below the curve.

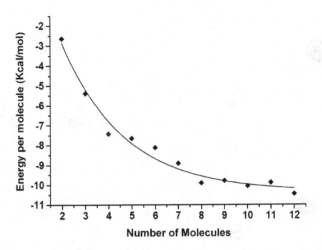

Fig. 5.8. A plot of the number of water molecules in the cluster *vs.* the energy per molecule for each cluster.

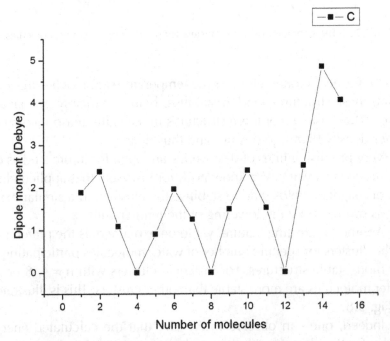

Fig. 5.9. This graph shows the variation of the dipole moment, μ, value with the number of water molecules in the cluster.

At some instance, the name of *magic numbers* was given to the n values of these specific cluster sizes. This is probably not an appropriate choice of nomenclature and may produce misleading interpretations.

Yet another pertinent observation is the variation of the dipole moment for the various clusters. The most stable ones, 4, 8, and 12 have values of $\mu = 0$ D (see Fig. 5.9). One can safely assume that this characteristic is due to the higher symmetry of these clusters.

The reader might be surprised that something as "simple" and familiar as water is today the object of very active scientific research.

CHAPTER 6

The Birth of Quantum Mechanics and Modern Chemistry

6.1. Preliminary Considerations

Between 1920 and 1930 a new theory emerged for the description of motion and forces for what we call microscopic systems such as electrons, protons and other subatomic particles. It received the name of quantum theory. This theory came to replace, for such a segment of the universe, the well-established classical mechanics, which adequately describe bodies such as butterflies or planets (we say adequately; it should not be understood as *easily* at all). The "old" mechanics had had solid phenomenological background since the eighteenth century, built by scientists such as Newton, Lagrange and Hamilton.

Almost a century has elapsed since the advent of quantum mechanics. But neither scientists nor the general public seem to continue to be as impressed by this "new" theory, as they were when it arose. To start with, elementary particles are not electrons and protons anymore, but quarks and leptons. We are testifying new revolutions and we have no way of knowing where we are to arrive. We are to control electronic devices and computers with our minds. Microscopic robots are to enter into our bodies to repair undesirable deviations in function or structure. Small bodies could be recreated at long distances from the original. Energy might be transported without a material support. We are creating living tissues and, maybe we are not far from the possibility of creating some kind of living being, if we dare.

The former is a brief list of examples of what can be done today or will be possible in the near future. The far future is beyond our prediction capability.

Even for the short-term ahead, futurology is an unsafe game, since drastic events such as natural and man-made disasters, which we have not learnt to predict and, even less to control, may occur. Among the last, wars are a preferred endeavour of man. The twentieth century had many, and two of them "deserved" the category of *World Wars*. The twenty-first century is already teeming with conflict and the threat of nuclear warfare, although apparently remote, is nonetheless ever present. It is claimed that Einstein said (so many things are attributed to Einstein): "I don't know with what weapons the third world war will be fought, but I'm sure that the fourth will be with arches and arrows".

6.2. Quantum Theory and Atomic Applications

Let us now focus on the subject of this chapter. One could describe quantum mechanics as the theory adequate for the study of the movement and interaction of microscopic particles.

Perhaps the first alert that classical theory would not serve for the microscopic range of magnitudes was brought in by the German physicist Max Planck in 1900. On the basis of his knowledge of thermodynamics and electrodynamics he suggested that energy could have only certain definite values, depending on the frequency of the light emitted from some sources (a *black body* in this case). As a result of his reasoning, Plank wrote the equation:

$$E = h\upsilon. \qquad (6.1)$$

In the above equation, E stands for energy and υ for frequency of the light (seg^{-1}). The number h received later the name of Plank's constant. What we learn from this is that energy may have only some definite, *discrete*, values. Energy is transported in pieces, i.e., *quanta* of energy.

A few years later, in 1905, Albert Einstein (Germany, 1879–USA, 1955), at that time in Germany, would apply the quantisation concept of Planck for the interpretation of the *photoelectric effect.* We know today that this effect is related to the ionisation potential of the metals studied by Einstein.

Some years afterwards, in 1911, Rutherford (Ernst Rutherford, New Zealand 1871–England 1937), who had moved from Cambridge to Montreal, presented his model for atomic structure. The model was consistent with all previous experimental knowledge on atomic systems. Thus, our present model of a positive nucleus and a cloud of negative electrons was established.

Such was the background available for the work of the Danish physicist Niels Bohr when he proposed in 1913 what we call the Bohr atomic model, presented in Chapter 2, Sec. 2.2.

In 1924, Louis de Broglie (duc de Broglie, France 1892–1987) in Paris wrote a thesis in which he treated the equations for an electron, as if it were simultaneously a particle and a wave. Soon afterwards (1925–1926), the Austrian physicist Erwin Schrödinger presented and discussed what is now known as the *wave equation* or, more often, the Schrödinger equation.

Various building blocks were incorporated into the theory in a short period of time until the days one could mark as the closure of the building of quantum mechanics: the uncertainty principle of Heisenberg (1927) and the probabilistic interpretation of the theory by Born (1926).

But at this same epoch we may arbitrarily recognise the emerging of what we now treat as a discipline of its own: quantum chemistry. Between 1925 and 1927, the Austrian-Swiss scientist Wolfang Pauli, as well as other authors, had recognised that electrons had a further property, that is, they *spin* and this spin admits only two values: up or down; clockwise or counterclockwise; $+\frac{1}{2}$ or $-\frac{1}{2}$ (a.u.). The spinning was at that time an useful image, but what we call spin of the electron (and other elementary particles, with values not necessarily equal to $\pm\frac{1}{2}$) is an intrinsic property of matter.

Pauli also formulated the *exclusion principle,* meaning that: a single energy level in an atom (or molecule) could be occupied at most by two electrons and that they ought to have different spins.

D.R. Hartree from England (1928) and V. Fock from Russia (1930) taught us how to write proper functions for many-electron atoms and J.C. Slater, in the USA, initiated in 1929 the theory for the treatment of complex atoms. These three authors and their physico-mathematical structures are ever present in our theory.

Let us emphasise how the pioneering work of Hartree, Fock, Slater and Pauli gave rise to the notion that the interactions of the electrons with the atomic nucleus was, in a first approximation, by far more important than the interaction among the electrons themselves. This hierarchy marked all the sequential progress in the understanding of atomic (and molecular) electronic structure and is evidenced in any text of chemistry.

6.3. The Applications to Molecules and Modern Quantum Chemistry

In the 1930s, quantum theory goes beyond the study of atomic structure and the first calculations for molecules are presented. Let us quote only the very simplest cases and just a few authors: the hydrogen molecule ion (H_2^+) discussed by Hylleraas (1931) and Jaffé (1934) and the hydrogen molecule calculated by Heitler and London in 1927. From the second half of the twentieth century, the applications of these new theories to molecular systems would permeate the literature with research articles, new journals and many books.

At this point it may be useful to give a practical definition of quantum chemistry as the application of quantum mechanics to molecular systems.

The constant methodological evolution and progress of quantum chemistry has a deep connection and, to a large extent, also mutual feedback, with computer technology. Since the 1960s, we have had a constant increase in memory, processing capacity and speed of computer machines. At the same time, available algorithms are permanently improved.

The tendency nowadays continues to be towards ever-increasing applications of quantum chemistry. As calculation procedures and capacity improve constantly, the systems being studied are larger every day; proteins, liquids, polymers are subjects of present calculation capability. For large systems, classical mechanics, statistical mechanics, mathematics and thermodynamics play a crucial role alongside quantum chemistry.

Quantum chemistry has provided us with mathematical tools for the calculation and prediction of properties of molecules, even for the design of new molecules with desirable properties. This is the case, for instance, for the planning of the synthesis of new drugs aimed at the treatment of specific diseases.

As we shall see in the following chapter on molecular orbital theory, quantum chemistry has also brought to us a renewed understanding of the chemical structure of molecules.

CHAPTER 7

Molecular Orbital Theory and the Recovery of Classical Chemical Notions

7.1. Preliminary Considerations

Theories and knowledge are certainly submitted to previous history as all our notions on the present world are. If some experiments or theoretical elaborations had occurred, say, ten years before, or ten years after they actually did occur, one may ask, whether such theory would have chosen the very same pathway that it did. Interesting simulations on such a theme could be devised.

Following the former line of thinking, one can also imagine that some experiments are designed to corroborate previous theoretical propositions. Certainly, in most cases, there is no intention to produce a fake result; but simply because the detector device, whatever it was, was not calibrated for a response other than the predicted.

Anyway, the proposition of Born (Max Born, Germany, 1882–1970, Britain) and Oppenheimer (Robert Oppenheimer, USA 1904–1967) in 1927 that the study of the motion of nuclei could be separated from the equations for the motion of electrons has marked practically the whole construction of quantum chemistry. Born and Oppenheimer based their formulation on the difference between the masses of electrons and nuclei. Indeed, the mass of a proton (the lightest nucleus is the hydrogen atom, the nucleus being a single proton) is almost 2,000 times larger than the mass of the electron.[1]

Thus, the lesson was: let us consider the nuclei as in fixed positions, while we calculate the electronic distribution in their surrounding and all the consequences we may draw from there. This worked marvelously and still works, although nowadays we know also how to deal with nuclear motions as well and how this movement affects various molecular properties.

7.2. The Independent Electron Approximation

In the 1930s, two scientists, Hartree (Douglas Hartree, England 1897–1958) and Fock (Vladimir Aleksandrovich Fock, Russia 1898–1974) developed a new formulation for the study of the electronic structure of atoms (equally valid for molecules indeed). Their approach avoided the very difficult mathematical intricacies of the solution of the equations of the electron-electron Coulomb type repulsion, E_{ee}:

$$E_{ee} = e^2/r_{ij}, \qquad (7.1)$$

Where e is the charge of an electron ($-e \times -e = + e^2$) and r_{ij} is the (variable) distance between electrons i and j. Instead, they considered that each electron would merely feel an average field produced by the rest of the negative particles. This assumption made mathematical solutions reasonably affordable and the results surprisingly close to the exact values. Thus, as we anticipated in the previous section, inter-electronic motion can be separated for each electron to a good approximation.

In such manner, the total energy, E, for an atom or molecule is separated as

$$E = E_0 + E_{ee}. \qquad (7.2)$$

In Eq. (7.2), the term E_0 contains the kinetic energy of the electrons, the potential energy attraction between electrons and nuclei and the *average* energy repulsion between the electrons. The Independent Electron Approximation consists in the resolution of the equations that lead to the value of E_0; this term is obtained at a relatively low computational cost and provides reasonable estimates for many properties.[2]

When we use the Independent Electron Approximation, the function which describes a system of n electrons, $\psi(1, 2, 3, \ldots, n)$, usually called *wave function*, may be written as a product of functions related to a single electron each:

$$\psi(1,2,3,\ldots,n) = \psi_k(1)\psi_l(2)\psi_m(3),\ldots,\psi_q(n) \qquad (7.3)$$

The one-electron functions, ψ_i, are called *atomic orbitals*, in the case of atoms, and *molecular orbitals*, in the case of molecules. The atomic orbitals ψ_i are the very same atomic orbitals ψ_{nlm} which we defined in section 5 of Chapter 2. The molecular orbitals are built from them as we show in Sec. 7.3. below. To each orbital, ψ_i, an energy value, ε_i, is associated, which comes to be the energy of the electron in that particular orbital.

Hence we have the association between the one-electron functions ψ_i and the one-electron energies ε_i:

$$\psi_k \rightarrow \varepsilon_k$$
$$\psi_l \rightarrow \varepsilon_l$$
$$\psi_m \rightarrow \varepsilon_m$$
$$\cdot$$
$$\cdot$$
$$\cdot$$
$$\psi_q \rightarrow \varepsilon_q$$

Not all one-electron energies need be different. In some cases, we may have $\varepsilon_r = \varepsilon_s$. When this situation arises, orbitals ψ_r and ψ_s are said to be (two-fold, for this example) *degenerate*. For instance, for the hydrogen atom, the three different orbitals *2p* for the various possible values of m ($m = -1, 0, +1$) have the same energy, they are three-fold degenerate. As commented in section 5 of Chapter 2, these three orbitals (let us call them as *2p_{-1}*, *2p_0*, and *2p_{+1}*), although represented by different functions, have the same energy and would reveal themselves experimentally as being different only if the atom was submitted to the action of an external magnetic field.

The latter considerations are equally valid for *3p, 4p, ...* orbitals, reminding us that the energy increases with the principal quantum number *n*. In general, the angular quantum number *l* leads to *2l + 1* degenerate functions.

The atomic orbitals *s, p, d, f, ...* are in fact the building blocks for the construction of the electronic structure of molecules (see Chapter 3). For practical geometrical purposes (since molecules adopt some specific form in space) the set p_{-1}, p_0, p_{+1} has been replaced in most applications by the equivalent set p_x, p_y, p_z. These latter functions have the great advantage of being aligned along the axes of a Cartesian co-ordinate system.

Similar geometrically adapted sets are adopted for *d, f,* and higher atomic orbitals.

As a result of the Independent Electron Approximation here discussed and of Pauli's Exclusion Principle (see Chapter 6), the electrons occupy the available energy levels, ε_i, by pairs, with opposite spin values, each level *i* being described by the atomic or molecular orbital function, ψ_i. We illustrate the electron occupancy for the ground state of a molecule in Fig. 7.1.

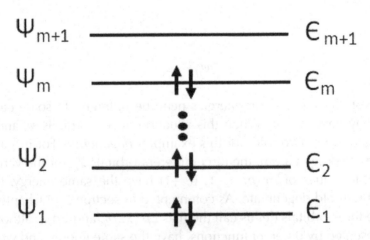

Fig. 7.1. The electron distribution for the ground state of a neutral molecule.

In the previous figure, the highest occupied molecular orbital (HOMO) is ψ_m, it is common to call it Ψ_{HOMO}; the lowest unoccupied molecular orbital (LUMO) is $\psi_{m+1} \equiv \Psi_{LUMO}$; this and higher empty orbitals are denominated as *virtual* orbitals. Arbitrarily, we classify the group Ψ_{HOMO-2}, Ψ_{HOMO-1}, Ψ_{HOMO}, Ψ_{LUMO}, and Ψ_{LUMO+1} as the set of *frontier orbitals*. The energies and shapes of these frontier orbitals have a crucial role for the properties, reactivity and ultraviolet and visible (UV/visible) spectra of the respective molecules.

7.3. The Mathematical Form of Molecular Orbitals and Graphical Visualisation

In the year of 1951, the scientists C.C.J. Roothaan (Dutch, 1918) and G.G. Hall (from Northern Ireland, 1925) offered in separate publications the proposal of writing the molecular orbitals as combinations of atomic orbitals. In such a way, for the molecular case, the various ψ_r functions of Eq. (7.3) are written as a *linear combination*[3] of atomic orbitals (either Slater type or Gaussian type):

$$\psi_r = \Sigma\, C_{r,a}\varphi_a \qquad (7.4)$$

In the former equation, the φ_a are atomic orbitals centered on the nucleus a and the $C_{r,a}$ are numerical expansion coefficients calculated as to form the **best** linear combination possible for the representation of a particular molecular orbital. The indexes r play the role of labeling the given one-electron function (r may be k, l, m,..., q) as in Eq. (7.3). Of course, on a given atom various atomic orbitals may be centered, depending on which is the particular atomic element involved.

We can clarify the format of molecular orbitals with a few examples. Let us consider, for instance, the case of methane, CH_4. Each H atom will contribute with a $1s$ function. The configuration of the C atom is $1s^2 2s^2 2p^2$. Thus we have the available functions 1s, 2s, $2p_x$, $2p_z$, and $2p_z$ on the carbon atom. Figure 7.2 below shows the numeration for the H atoms.

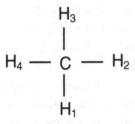

Fig. 7.2.   The numbering of the H atoms in the methane molecule.

A particular molecular orbital for methane, say ψ_k, will have then the analytical expression

$$\psi_k = C_{k,1}\ 1s(H_1) + C_{k,2}\ 1s(H_2) + C_{k,3}\ 1s(H_3) + C_{k,4}\ 1s(H_4)$$
$$+ C_{k,5}\ 1s(C) + C_{k,6}\ 2s(C) + C_{k,7}\ 2p_x(C) + C_{k,8}\ 2p_y(C)$$
$$+ C_{k,9}\ 2p_z(C) \tag{7.5}$$

The molecule of methane has ten electrons and, according to Pauli's exclusion principle (see Chapters 2 and 6), the ground state will have five occupied molecular orbitals, as shown in Fig. 7.3.

As one can see in the figure, the highest molecular orbital is actually triply degenerate, and all three are HOMO's with equal rights.

The explicit forms for the first five molecular orbitals of methane are:

$$\psi_1 = 1s(C) \tag{7.6}$$

$$\psi_2 = 0.70\ 2s(C) + 0.20[1s(H_1) + 1s(H_2) + 1s(H_3) + 1s(H_4)] \tag{7.7}$$

$$\psi_{3(HOMO)} = 0.60\ 2p_y + 0.30[1s(H_1) - 1s(H_2) + 1s(H_3) - 1s(H_4)] \tag{7.8}$$

$$\psi_{4(HOMO)} = 0.60\ 2p_y + 0.30[1s(H_1) + 1s(H_2) - 1s(H_3) - 1s(H_4)] \tag{7.9}$$

$$\psi_{5(HOMO)} = 0.60\ 2p_y + 0.30[1s(H_1) - 1s(H_2) - 1s(H_3) + 1s(H_4)] \tag{7.10}$$

The first unoccupied molecular orbital is also a triple degenerate set ($\psi_{6(LUMO)}$, $\psi_{7(LUMO)}$ and $\psi_{8(LUMO)}$) with a $2p_z$ orbital on the C atom

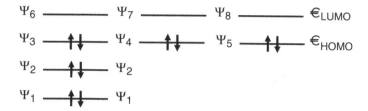

Fig. 7.3. Illustration of the electronic distribution for the ground state of the methane molecule showing the first five occupied molecular orbitals. The first empty (virtual) energy level, LUMO, is also indicated.

and similar combinations for the 1s orbitals on the H atoms as for the HOMO. We must emphasise again that for degenerate orbitals, the forms of the orbitals are different but the energies are strictly identical.

We may describe the orbitals in Eqs. (7.6) to (7.10) as follows:

ψ_1, with the lowest energy, is an isolated full 1s orbital centered on the C atom, with no contribution from the H atoms.

ψ_2 is a bonding molecular orbital mixing a 2s atomic orbital on carbon with a symmetric linear combination between the 1s orbitals on the hydrogen atoms.

The three degenerate HOMO's, $\psi_{3(HOMO)}$, $\psi_{4(HOMO)}$, and $\psi_{5(HOMO)}$, employ a $2p_y$ atomic orbital on carbon with three different non-symmetrical linear combinations of 1s atomic orbitals on the hydrogen atoms.

The three degenerate LUMO's have a similar form as the HOMO's, but employing a $2p_z$ atomic orbital on the carbon atom.

It is a common practice to visualise the characteristics of molecular orbitals graphically rather than through analytical expressions as Eq. (7.6.) through (7.10.). Figure 7.4. thus shows the design of the methane molecular orbitals as provided by the computer employing the appropriate program. Several programmes are available for the preparation of these graphs; they are either part of the calculation routine or may be called separately.

The graphical representation of the molecular orbitals is an indication of how the density of the electrons is distributed in

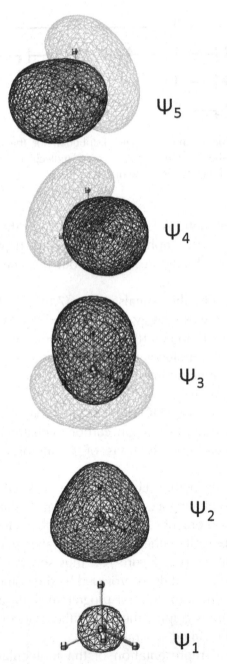

Fig. 7.4. Graphical representation of the molecular orbitals of methane.

the framework of the molecular skeleton for each one-electron function ψ_i. Since a molecular orbital may have positive and negative portions, colours are often used to distinguish them or, simply, black for the positive and white or gray for the negative as in this chapter.

Let us provide one more example of molecular orbitals, using a small molecule again, carbon monoxide, CO. The electronic configuration of oxygen is $1s^2 2s^2 2p^4$. Thus we have the same type of atomic orbitals as for carbon. We should remember, however, that the constants characterising the functions for oxygen will have values different from those for carbon (see Chapter 2, Sec. 2.5). For the CO case we shall have 14 electrons. The frontier molecular orbitals are represented in Fig. 7.5.

The reader may anticipate how this mathematical description of molecular orbitals becomes more and more cumbersome as the size of the molecule increases and heavier atoms are included.

In such a way, when the large and efficient computer machines started to be commercially available in the 1960s, quantum chemistry made a simultaneous advance. Ever since, there was a constant mutual stimulus between quantum chemistry and computer progress in hardware and software. Indeed, molecular calculations was one of the areas in which there arose a heavy demand for computation capacity, as did finances, meteorology and other subjects. Nowadays, intelligent informatics is present in every activity: medicine, communications, religion, leisure, etc.

Computer technology has seen constant progress in miniaturisation, algorithms and processing speed. In fact, there are two competing main streams of development: large frame machines and personal computers (PCs). Each solution has its applications and users. The limit has not been reached. One can say that the computation capacity of a main frame today will be available on a PC in some five years from now.

As for the case of quantum chemistry, PCs seem to be dominating the scene; clusters of PCs also became very popular.

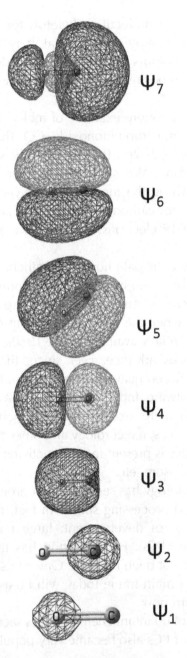

Fig. 7.5.   Graphical representation of the molecular orbitals of the CO molecule.

7.4. Molecular Orbitals and the Structure and Properties of Molecules

All quantum chemical calculation procedures have Eq. (7.3), i.e., the wave function, as the starting point. The degrees of sophistication of the method chosen for a calculation depends of the accuracy desired (including eventually *exact* values) and the property of interest. Any property which can be measured may be calculated as well; some useful "descriptors" can be calculated but are not directly observed as for instance bond orders and atomic charges described below. Of course, once the scope of the calculation is decided, the appropriate method and programme should be selected. The machine adequate for such calculations should also be available, either in the researcher's laboratory, or attainable by remote access.

The methods of quantum chemistry are thus employed in conjunction with the analytical form for the molecular orbitals (Eq. (7.4)) to provide theoretically various molecular properties that describe the chemical and physical behavior of substances. In some cases, we may have measurements to compare with, in other situations, we ought to relay solely in the theoretical information.

Now we list the main properties (we may say descriptors) that can be calculated for a molecule.

7.5. Molecular Geometrical Structure

This means the spatial arrangement of the atoms of a molecule, vicinities, atom-atom distances, angles and dihedral angles. In all cases, we will choose as the most stable that geometry conformation with the lowest total energy. Notwithstanding, other geometries, with higher energy values, are also possible, these corresponding to excited states of the molecule.

We ought to recognise that atoms in a molecule are never "still". Even a diatomic molecule, say H_2, will *vibrate*, that is, the two H atoms will move back and forth from each other many times per second. This movement is called vibration; it is of high frequency, that is, for instance for H_2, 12.47×10^{13} vibrations per second. Actually, the two hydrogen atoms never separate too far from each other, but they just vibrate slightly around a position which we call the *equilibrium distance* between the two atoms and which value is accepted as the H–H inter-atomic distance in the H_2 molecule. This value is of 0.74×10^{-10} m.

As the number of atoms in the molecule increases, more complex movements between the atoms may be visualised. For instance for ethane, H_3C-CH_3, the two methyl fragments ($-CH_3$) may rotate one respect to the other, passing across two specific positions called *eclipsed* and *staggered*, which we illustrate in Fig. 7.6.

In Fig. 7.6 we focus on two possible geometries of high symmetry. In one case the H atoms of the first carbon atom will stand exactly in front of the H atoms of the second carbon atom, so the last will be "hidden"; we call this configuration "eclipsed". In the other case the H atoms belonging to the second carbon atom are shifted in 30° with respect to the front atoms, thus we can "see" them: this is called staggered configuration. The figure also shows the evolution of the total energy, E, of ethane during the rotation. Both configurations exist naturally and are separated by a very low energy barrier; thus the methyl groups of ethane will be spinning almost freely.

In principle, the determination of the geometry of a small molecule should be a relatively simple task; yet, it is not necessarily that simple. For instance, the three-atom molecule ozone, O_3, is present in the earth atmosphere and its absence (believed to be provoked by man-made gas emissions) exposes the earth surface to undesirable solar radiations. The ground state of ozone has an open O–O–O form, with an O–O–O angle of *ca.* 120°. Ozone may also exist as a low-lying excited state in an equilateral triangle (all angles being 60°) geometrical form. The energies of both forms are extremely close, so that accurate calculation methods are required to decide if the first form has the lower energy.

As the number of atoms in a molecule increases, the number of variables (distances and angles) increases even more rapidly. For a diatomic molecule, say H_2, one has only the inter-atomic distance as a variable (see Fig. 7.7). In the case of a tri-atomic molecule as ozone (see above) one has the O–O distance and the O–O–O angle as variables; in the case of a tri-atomic molecule as, for instance, hydrogen

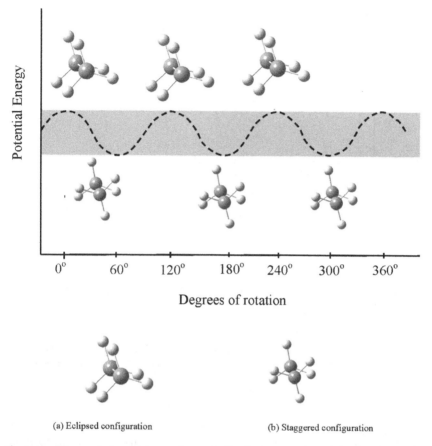

(a) Eclipsed configuration   (b) Staggered configuration

Fig. 7.6. The evolution of the total energy, E, of ethane as the two methyl segments rotate one respect the other.

cyanide, HCN, there are two distances to be calculated, d_{HC} and d_{CN} and the H–C–N angle.

For molecules of four or more atoms, we need to specify also the *dihedral angles*. Any three points describe a plane. When there are more than three atoms, the molecule can still be planar. This is the case, for instance, of the benzene molecule, C_6H_6 (Fig. 7.8a). All atoms are in the same plane (setting aside small vibrational instant deformations) and the carbon skeleton of the molecule has the form of a regular hexagon.

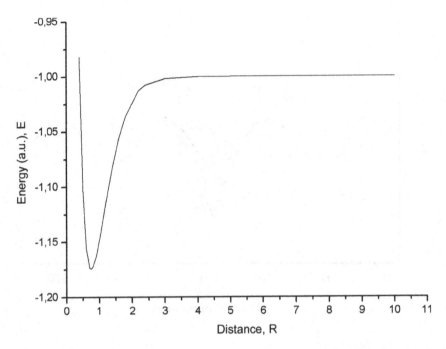

Fig. 7.7. The energy of the H_2 molecule as a function of the d_{HH} distance. R (0.074 nm) indicates the equilibrium distance between the two H atoms in the molecule.

On the other hand, some molecules with only four atoms will be non-planar. We can illustrate the need for the definition of dihedral angles with the familiar molecule of hydrogen peroxide, H_2O_2, the oxygenated water used as an antiseptic for small skin cuts. We place the O–O–H segment in a plane (see Fig. 7.8b). Then the second H atom will form a dihedral angle of 111.5° with that plane.

As molecules become larger and larger (hormones, vitamins, polymers, proteins, etc.; the list of classes of compounds is long), the computational problem becomes more and more demanding. The number of variables (distances, angles) to be simultaneously optimised is large. Almost nothing can be done without computers; even the simplest R_{HH} equilibrium distance for the hydrogen molecule will resist an *exact* solution by hand in reasonable time.

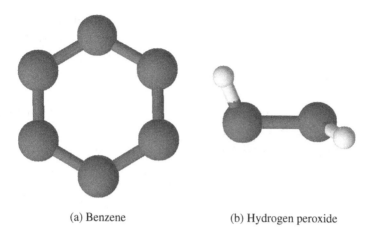

(a) Benzene   (b) Hydrogen peroxide

Fig. 7.8. (a) The benzene molecule; and (b) The H_2O_2 molecule showing the angle (111.5°) between the O–O–H plane and the O–H link out of plane.

In actual calculations, the user should avoid demanding from the computer the useless effort of searching for already known structures. In such a way, if we are interested in a derivative of the benzene molecule, the computer should be put to work for the unknown fragment, but not to redundantly again calculate the well-known benzene geometry.

Also, one should bear in mind that for very large molecules (say, several thousand atoms) many very close geometries are possible. Since the Boltzmann law determines the distribution of the molecules among the possible states, according to the temperature, several states of similar geometry will exist simultaneously.

Furthermore, often molecules are "soft"; some proteins in the human body, at 37°C and surrounded by many water molecules, may be constantly changing form, like a snake in water.

What we have described so far on the geometric structure of molecules, refers mainly to isolated molecules, some like to use the term "gas phase" molecule. The theory may also attack the challenge to calculate the structure of molecules in a solid or in solution.

To close this section on molecular structure, let us say that structure may exist where it is not expected to. Some years ago, our notion of liquid water was of isolated molecules, moving randomly at speeds determined by the temperature and colliding frequently between them. Now we know that water molecules organise themselves as clusters of water molecules of various sizes.

Since water is omnipresent in our lives, we have much to gain in understanding this ordering. For this purpose, we have discussed in some detail the structure of water in Chapter 5, "Inorganic Molecules".

7.6. Total Energy

The total energy of molecules varies with the number of atoms and the constituant elements. The larger the molecule, the larger (in absolute values) the energy.

Therefore, this property is most of the time employed for the comparison of similar structures like *isomers* (the isomers have the same atoms but different spatial arrangement). For instance, in the scheme below, n-butane and methyl propane are isomers (C_6H_{10}), but their total energy differs by 0.003 a.u. Thus, methyl propane is slightly more stable than n-butane.

$$CH_3-CH_2-CH_2-CH_3 \quad \text{Total energy} \quad -157.912 \quad \text{a.u.}$$
n-butane

$$CH_3$$
$$|$$
$$CH_3-CH-CH_3 \quad \text{Total energy} \quad -157.915 \quad \text{a.u.}$$
Methyl propane

Many other applications for total energy are possible. Figure 7.6 above shows the use of total energy to discuss the rotation of the methyl moieties in ethane. Also, Fig. 7.7. illustrates the employ of the total energy for the study of the stability of a diatomic molecule.

7.7. Standard Heat of Formation

The standard heat of formation is a thermodynamic property related to the total energy of a system which provides accurate information on the stability of molecules. In a chemical transformation:

$$Reactants \rightarrow Products,$$

heat may be produced by the reaction process or heat has to be provided in order for the reaction to occur. This heat is called *enthalpy*, symbol H, so we have

$$H_{products} - H_{reactants} = \Delta H \tag{7.11}$$

When $\Delta H < 0$, heat is liberated by the reaction, the products are more stable than the reactants. Such reactions are called *exothermic*, since heat is liberated.

If $\Delta H > 0$, heat has to be provided, the reaction is called *endothermic*, that is, heat from external sources is consumed.

In the particular case in which a molecule is formed from its constituent atoms, we refer to the enthalpy as heat of *formation*, symbol ΔH_f. Let us use water as a simple example:

$$H_{2(g)} + 1/2O_{2(g)} \rightarrow H_2O_{(g)} \tag{7.12}$$

where we denote with the symbol (*g*) that all three species are in the gas phase. Furthermore, for the better understanding of the stability of different molecules, we compare their heat of formation, ΔH_f, measured at 25°C in all cases. $\Delta H°_f$ is then denominated as *standard heat of formation*.

The $\Delta H°_f$ for water is −241.8 kJ/mol. Thus, H_2O is a stable molecule, which we know from our everyday experience; we can boil, freeze, overheat water and it will still be H_2O.

Standard heats of formation, as provided by modern computational algorithms, are useful variables for the discussion of molecular properties and transformations.

As for chemical transformation itself, we shall discuss this matter in more detail in Chapter 9.

7.8. Bond Strengths

The property of *bond strength* is often also called *bond order* and it is a quantitative measure of the strengths or binding force between two atoms (generally, but not necessarily, neighbours) in a molecule. In molecular orbital theory this parameter recovers classical notions on bonding, as single bonds, double bonds, triple bonds, aromatic bonds and hydrogen bonds, principally.

In Fig. 7.9. we show the values, as calculated, for the C–C bond in ethane, ethylene and acetylene. We also show benzene, in which double and single CC bonds should alternate but, as symmetry prevents the choice of which atom is which, there is 1½ bond between each pair of carbon atoms. The actual units for bond orders are atomic units, thus, the value 1.0, for instance, indicates that we have the density equivalent to the charge of one electron between the two respective atoms.

In Fig. 7.9(e) we show the calculated bond orders for the $S_3N_3^-$ ion; a classical description of the bonding in this interesting non-carbon aromatic ring is not obvious.

One can observe in Fig. 7.9 that the results of molecular orbital theory recover numerically (a) single, (b) double, (c) triple and (d) aromatic CC bonds, respectively. In Fig. 7.9e the ion, $S_3N_3^-$ is represented with the calculated SN bonds strengths of some 0.52 electron density. The trisulphur trinitride ion has a hexagonal geometry similar to benzene, and it is amenable to synthesis, although rather unstable by shock or heat.

Thus, molecular orbital theory, other than having the academic merit of recovering our classical notions on bonding, can provide interpretation and understanding for the enormous number of novel chemical families, which began to appear in the last century. Often, in the past, these new compounds were assumed as not viable, as were, for instance, the many compounds made from *noble* gas atoms.

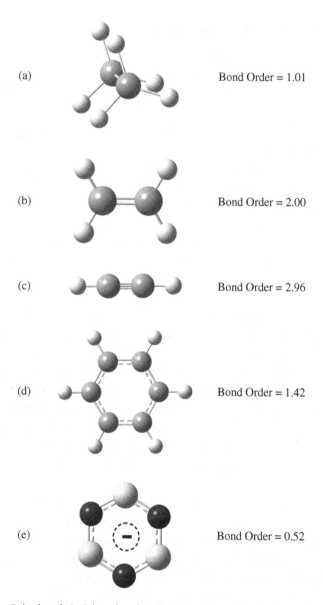

(a)   Bond Order = 1.01

(b)   Bond Order = 2.00

(c)   Bond Order = 2.96

(d)   Bond Order = 1.42

(e)   Bond Order = 0.52

Fig. 7.9. Calculated C–C bond orders for (a) ethane, (b) ethylene, (c) acetylene, (d) benzene, and (e) S–N bond order for the $S_3N_3^-$ ion. The respective classical chemical bonds are drawn in the figure.

7.9. Charge Density on Atoms

The estimate of the charge density surrounding an atom within a molecule is a very important contribution of the molecular orbital theory to the interpretation of the structure and properties of molecules.

Chemists have known for centuries that some atoms tend to attract negative functional groups or moieties more strongly than others. To give an extreme example, in the case of sodium chloride (NaCl) in a crystal or in solution (see Chapter 2, Sec. 2.6), chlorine will carry a full extra electron, forming the Cl^- ion. This is the case denominated as an *ionic bond* between Na^+ and Cl^-.

But charge can be unevenly distributed with the atoms still linked through what we call *covalent bonds*, that is, electrons shared between two nuclei holding the molecule together. Covalent bonds were described in Chapter 3.

It was the American chemist Linus Pauling (1901–1994)[4] who introduced the concept of the *electronegativity* of atoms, proposing a scale of values in 1932. Halogens are the most electronegative atoms, while alkaline metals are less so (actually, these metals tend to lose electrons in a combination). In spite of its great utility in predicting or explaining molecular formation and properties, electronegativity values are not amenable to calculation. At this point appears one more contribution of modern molecular orbital theory, allowing the numerical calibration of what is called the charge or electron density of an atom in a molecule.

All calculation routines have the capacity of providing the charge distribution on the atoms within a molecule. For a homonuclear diatomic molecule, nitrogen molecule for instance, N_2, the electrons will necessarily have the same concentration on each N atom, but for PCl, with different electronegativities for the atoms, the charge distribution will not be homogeneous (chlorine is slightly more electronegative than phosphorus). In the case of water, some extra negative charge will surround the oxygen atom. The former examples are illustrated in Fig. 7.10.

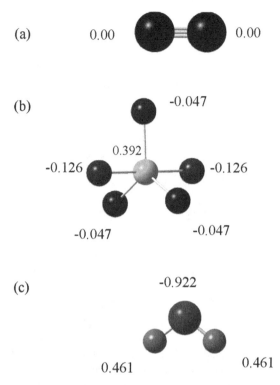

Fig. 7.10. Calculated charge distributions for (a) N_2, (b) PCl_5, and (c) water.

The atomic charge distribution in a molecule also depends on spatial distribution other than electronegativity. For a highly symmetrical molecule with identical atoms (see N_2 in Fig. 7.10) charges are necessarily identical (zero in this case). But even for a symmetrical molecule such as benzene (see Fig. 7.11) the slight electronegativity difference between carbon and hydrogen atoms will produce small negative charges on the former and the equivalent positive charges on the hydrogen atoms.

On the other hand, the decrease of symmetry is also at the origin of charge differentiation. So in Fig. 7.11(a) all six carbon atoms and the six hydrogen atoms are equivalent, thus all carbon atoms in benzene have identical charges; the same holds for the hydrogen atoms.

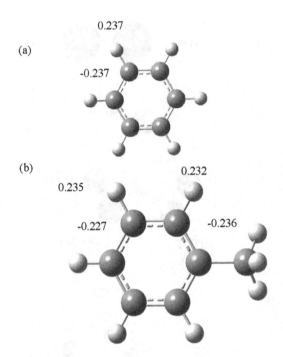

Fig. 7.11. Charge distribution for (a) benzene and (b) toluene. In the case (a) the slight charge differences between the hydrogen atoms and the carbon atoms for benzene are shown. In case (b), only the carbon charges on the phenyl ring are shown.

Yet, for the case of toluene, Fig. 7.11(b), the methyl group cuts the molecule by an imaginary plane perpendicular to the phenyl plane. Thus, we have only two pairs of identical carbon atoms (those in positions *ortho* and *metha,* or carbon atom number 2 with carbon atom number 6 and 3 with 5).

CHAPTER 8

The Nature of the Atomic Nucleus

8.1. Preliminary Considerations

Nowadays we believe to have a more or less valid image of the electronic structure of an atom. The nucleus has a positive charge density confined within a small volume and the electrons form a negative cloud around the nucleus. Nothing guarantees that someday we will not be forced to change this model as we go deeper into the nature of matter.

In comparison, what we know on the constitution of the nucleus itself is much less conclusive. Suffice to say that presently a gigantic experiment is underway in Europe (there is a similar project taking place at the Fermilab in Illinois, USA) to try to find what physicists believe would be the ultimate piece (the hypothetical particle known as the Higgs boson) that would be the key to understanding the constitution of the nucleus. The experiment has been prepared for years at CERN (Conseil Européen pour la Recherche Nucleare), with a particle accelerator underground at the border between Switzerland and France, called the Large Hadron Collider (LHC).

We may wonder whether the former experiment will really provide that final answer, or will nuclear nature continue to be elusive?

In what follows we present a very brief description of what is presently known on the constitution of the nucleus in accordance with the aim of this book.

8.2. The Rutherford Experiment

A piece of the element polonium (Po) was placed inside a block of lead; the polonium was known to emit alpha particles. Alpha particles are He atom nuclei. Hence, having the nuclear charge of He Z = 2 and lost the two electrons, the nucleus is charged positively. The beam emitted by the element was let outside the block through a hole and directed towards a thin foil of gold. A circular detector of zinc sulphide (ZnS) was placed; this material would scintillate with the impact of the alpha particles.

Most of the particles were transmitted unaffected through the gold sheet, some were deviated at varying angles, and a few were reflected.

This experiment was designed by Ernest Rutherford and his collaborators in 1911. Rutherford was aware of a previous suggestion by the Japanese physicist Hantaro Nagaoka (Japan 1865–1950) in 1904 and interpreted the fact that the larger number of alpha particles passed through the foil unperturbed and very few were reflected was because most of the atom was empty space. The mass of the atoms was therefore concentrated in a small region (the nucleus).

The fact that some of the incoming particles were deviated induced him to believe that the nucleus was positive, thus repelling the positively charged alpha particles.

In a similar experiment in 1918, Rutherford put forward the proposition of the existence of the proton, with charge of equal value as the electron, but with opposite sign. In 1932, James Chadwick (England 1891–1974) provided proof for the presence of yet another particle in the nucleus: the neutron, with mass similar to the mass of the proton and charge equal to zero. In Fig. 8.1 we show the atomic model proposed by Rutherford.

8.3. The Fundamental Forces

Soon it became puzzling that the nucleus could be stable in spite of the repulsion between the positively charged protons. Gravitation was too weak for the masses involved and electromagnetic forces

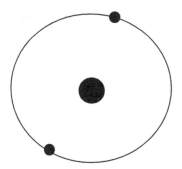

Fig. 8.1. The atomic model of Rutherford for the He atom.

Table 8.1. The four forces in nature

Force	Mediating particle	Main action in nature
Electromagnetic force	Photon	Stability of atoms chemical structure
Gravitational force	Graviton (never detected)	Solar system galaxies
Strong force	Gluon	Nuclear stability
Weak force	Heavy and short living particles (bosons W)	Nuclear radioactivity solar activity

would not explain the tight binding of uncharged neutrons inside the nucleus.

Thus, a new force was introduced in science: the *strong force*. This is the force responsible for nuclear stability. The theory which nowadays studies this type of structures and interactions is called quantum chromodynamics. The particles which interact through the strong force receive the name of hadrons while other particles are called leptons, the main exponent of the last being the electron.

Electrons do participate in other types of interactions, namely, the *weak force* and the *electromagnetic* interaction. These interactions are mediated by the interchange of virtual particles constantly. In such way, protons and neutrons would stay linked by the interchange of *gluons* (name derived from the word glue).

In Table 8.1 we summarise our present understanding of the various forces in nature.

8.4. A Closer Look into the Atomic Nucleus

Technological progress, mainly in particle accelerators, has led physicists to wonder whether protons and neutrons were the fundamental particles in nuclei. The indivisibility of protons and neutrons was put in doubt and, since 1964, the *quark* model gained credibility.

Quarks would thus be the particles that form protons and neutrons. Quarks interact through all the four forces described in section 8.3. above and have non-unitary charges, with values of either −1/3 or +2/3. Quarks have *properties* which are called *flavours*. There are six flavours with the denominations up, down, charm, strange, top and bottom. Two quarks up (charge +2/3) and one quark down (charge −1/3) form the proton while the neutron is formed by two quarks down and one quark up. Quarks with the other properties (charm, strange, top and bottom) may only be artificially produced in accelerators.

Quarks also have +1/2 or −1/2 spin values, the same as electrons, and thus they obey Pauli's exclusion principle (see Chapter 6). Experiments detected a particle with three quarks with the same flavour. This particle is denominated Ω^- (omega-minus), which would necessarily have a $\alpha\alpha\beta$ or a $\alpha\beta\beta$ spin arrangement, thus violating Pauli's principle. As a result, physicists postulated yet a new property: *colour*. So quarks could be red, green or blue.

Almost needless to say, neither *flavour* nor *colour* in this context have any relation whatsoever with our notion of their common meaning. Are quarks the smallest particles? Why should they be?

8.5. Radioactivity

The word *radioactivity* embodies both the emission of particles or energy (photons) by the nucleus. Nuclei which are unstable in nature are said to *decay*, emitting radiation and transforming into more stable nuclei. The particles which may be *emitted* are:

i) alpha (α) particles formed by a He atom nucleus (two protons and two neutrons);

ii) electrons (β⁻ particles); and

iii) positrons (β⁺ particles), that is, particles with the mass of an electron and positive unitary charge; the positron is the antiparticle of the electron.

When an atom is excited to a higher electronic state and then falls back to a lower level, energy is emitted in the form of photons. A similar phenomenon occurs for nuclei in excited states. When the nucleus returns to its lower energy state it emits photons, named γ-rays (gamma rays). The energy of γ-rays is many orders of magnitude larger than it is for electronic emission and may produce severe damage in the deoxyribonucleic acid (DNA) which carries our genetic information. More information of the role of DNA in inheritance may be found in Chapter 11.

8.6. Radioisotopes

It is first necessary to define the term *isotope*. Every element (atom) has a definite location in the periodic table. The chemical properties of an atom are a result of the number of protons, which equals the number of electrons (we refer to *neutral* atoms indeed). But the number of neutrons in the nucleus may vary, not affecting the chemical nature of the element.[1]

Thus, isotope literally means *same place* in ancient Greek, i.e., same box in the periodic table, the same element with a varying number of neutrons in the nucleus. The very familiar example of the carbon atom shall better illustrate this meaning. The most common carbon isotope is $^{12}C_6$, with 6 protons and 6 neutrons. In Table 8.2. we show the three isotopes which constitute the carbon found in nature.

Table 8.2.   The isotopes of carbon

Common name	Symbol	Abundance (%)
Carbon-12	$^{12}C_6$	99
Carbon-13	$^{13}C_6$	1
Carbon-14	$^{14}C_6$	One part in 10^{12}

Of the three isotopes of carbon, only carbon-14 is radioactive, which has scientific and technological utility, as shown below.

In general, elements in nature are a mixture of several isotopes, some of which are radioactive. Thus, the human body is slightly radioactive. The emitted radiation may consist of α, β or γ radiation.

8.7. Applications of Radioactivity

A few applications of radioactivity are described in this section. It is important to remark that in the following techniques, the irradiation of living bodies or materials does not induce radioactivity in them.

8.7.1. *Applications in medicine*

The radio-isotope of an element shall follow the same metabolic route as the natural substance in animals or plants. The iodide 131 isotope is absorbed mainly by the thyroid gland, as is iodide 127, the stable isotope (iodide has 37 isotopes). Iodide 131 emits β particles and γ-rays and has a *half-life* (half-life of an isotope is the time required for half of the specific nucleus to undergo radioactive decay) of eight days. A detector placed in front of the thyroid gland will measure the amount of iodide absorbed and consequently whether there is an appropriate or deficient absorption rate by the organism.

Yet in another application, radioactive sources are employed to destroy cancer affected cells in human bodies. There are several ways in which radiation for therapy can be delivered, i.e., external beam radiation (for many years the *cobalt bomb* has been used for that purpose), internally placed radiation devices and radiotherapy through chemotherapy.

Radioactive sources may also be used for the sterilisation of chirurgical instruments and materials.

8.7.2. *Applications in agriculture*

Very minimal amounts of radioactive elements — trace elements — may be used to follow the metabolism of plants, the absorption of nutrients, distribution in roots and leaves.

Another application is to sterilise males of harmful insects with γ-rays and let them loose. These sterile males will compete with normal specimens and reduce reproduction.

Pest insects marked with radio-isotopes will allow the identifying of their natural predators. The predators may be then brought into the battle against the pest, without the use of insecticides, in the hope that the predator does not become a plague.

Irradiated food commodities may be stored for more than one year without loss of nutritional and visual properties.

8.7.3. *Dating of fossils through C-14 content*

Carbon-14 is produced in the high atmosphere by the action of cosmic rays on nitrogen. This radioactive isotope combines with oxygen in the same manner as carbon-12 to form CO_2. The carbon dioxide is absorbed by plants, so that its presence may be detected in vegetable and animal fossils. Carbon-14 decays through the reaction

$$^{14}C_6 \rightarrow {}^{14}N_7 + {}^{0}e_{-1} \text{ plus energy,} \qquad (8.1)$$

with half-life of 5 600 years. Measuring the radioactivity in such samples it can be inferred when the absorption of CO_2 stopped, that is, the end of that particular life. The amount of carbon-14 left allows the dating of the sample.

8.8. The Evil Inside the Nucleus

While discussing non-chemical energy sources in Chapter 10 we necessarily mention the nuclear energy, controlled nuclear reactions which produce heat, and eventually mechanical and electrical energy (Sec. 10.4). Yet, the enormous energy content in the nuclear mass may also lead to nuclear weapons. The equivalence between mass and energy was formulated in its present form by Einstein in 1905:

$$E = mc^2; \qquad (8.2)$$

E is energy, c is the speed of light and m represents the mass. For an idea of what this equivalence represents, let us say that the monstrous destruction caused by the atomic bomb dropped over Nagasaki on August 9 of 1945 represented the loss of approximately the mass of 1 g of Plutonium. From formula (8.2) one obtains that 1 g of mass is equivalent to the explosive power of 21 kt of TNT = 10^9 g of TNT.

TNT is an abbreviation for 2, 4, 6 trinitrotoluene, a very powerful chemical explosive. Figure 8.2 shows the chemical formula of TNT.

The atomic bomb's power is based on *nuclear fission*. The fission produces at the same time neutrons which collide with other atoms and produce a *chain reaction*. Energy is produced in the form of light, heat, γ-rays, β particles, etc. The hot material produced after the explosion (millions of degrees Celsius) ascends quickly and then expands, giving rise to the infamous mushroom cloud.

Another process, called *nuclear fusion* led to the hydrogen bomb, many times more destructive than the atomic bomb. In the case of fusion, light atoms such as hydrogen collide and melt together,

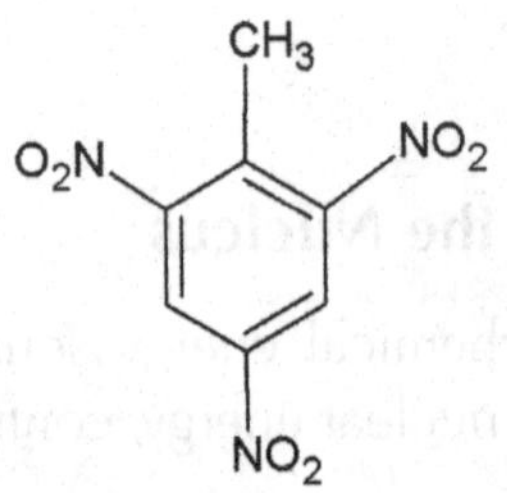

Fig. 8.2.   The chemical structure of the explosive 2, 4, 6 trinitrotoluene, TNT.

creating heavier atoms such as deuterium or helium and liberating an enormous amount of energy. To overcome the nuclear repulsion, an atomic bomb is used to trigger the hydrogen bomb!

After the end of World War II, many nuclear weapons have been produced and stored, but, so far, never used again. Whatever superpower was tempted to launch a nuclear attack was deterred by the certainty of reciprocal destruction. But nowadays, nuclear weapons are in many hands ...

Let us end this chapter by recognising that it is nuclear fusion in *our* sun that provides the energy to warm the earth.

CHAPTER 9

Chemical Transformations and Reactions: Velocity and Energy Balance

9.1. Preliminary Considerations

In order to sustain their activities, living beings need to obtain their energy from some source. The sun is our main source of energy and some organisms may obtain its energy directly from it. This process is called photosynthesis and we shall come back to this subject in Sec. 10.7. of the following Chapter. Chapter 10 deals with various forms and sources of energy.

In the photosynthetic process, light energy is transformed into chemical energy. The organisms capable of this transformation possess a molecule called *chlorophyll* which *catalyses* this reaction. Such beings are denominated *autotrophs* and include the plants, algae and some bacteria. Very exceptional autotrophs, which do not derive their energy from the sun, do exist, although rarely; we mention an example in footnote (1) of Sec. 10.2.

The following equation (same as (10.7) in Sec. 10.6) represents the photosynthetic process:

$$6H_2O + 6CO_2 + light \rightarrow 6O_2 + C_6H_{12}O_6 \qquad (9.1)$$

The hydrocarbons so formed, as $C_6H_{12}O_6$, a sugar molecule, are part of the energy reservoir for further use.

The other big group of living beings, humans included, are *heterotrophs*; they are not capable of photosynthesis, so they obtain the energy they require through food. And indeed, they require energy for a plethora of processes, as, for instance, locomotion, work, fighting infections, maintaining body temperature, growing, reproduction, and so on.

All the above energy transformation processes and others we shall examine further, obey the laws of thermodynamics. Although the initial concepts and worries on this subject were motivated by attempting to make efficient machines and processes, our above motivating words based on living systems are justified. Indeed, for most of us, our daily energy input is much more familiar than that of complex machines.

9.2. Basic Notions

Thermodynamics divides the *universe*[1] for functional purposes, into two parts: the *system* and the *surroundings*. The system is where we perform our experiments and the surroundings represent the rest of the universe.

Regarding the frontier between our system and the surroundings, we classify the system in various categories. To start with, the system may be *open*, *closed*, or *isolated*.

In the case of an open system, the frontier permits the transfer of matter and energy between the system and the surroundings. The human body is an example of an open system, breathes and expels gases, eats, drinks and produces solid and liquid residues. As for energy, it has a permanent interchange of heat with the environment.

A closed system prevents the passage of matter but allows the transfer of energy. One can think of a transparent box of a very strong material, which prevents the passage of all solid matter but allows the entry of radiation and heat.

Finally, there are the isolated systems. In this case there is neither transfer of matter nor energy, nor work. A good example is an insulated rigid container, such as an insulated gas cylinder.

9.3. Energy

Let us now discuss *energy*, a good starting point for those interested in chemical transformations and how thermodynamics influences them. Energy is the capability of a system to perform *work*. Work (w) is the transfer of energy through the ordered movement of molecules. When work is performed on a system, the system gains in capacity in performing further work, thus we say w > 0. It means that the energy of the system increased. When the system performs work, its capacity to perform further work is diminished, thus we say w < 0 and the energy decreases.

If we compress a spring, we perform work on it and it accumulates energy to restore its initial form when we end our pressure.

Alternatively to the transfer of energy through the ordered movements of molecules (work), it may be transferred in a disordered manner, which we consider as *heat* (q) transfer. An experience common to almost everybody is that of a pot of water being heated and we can visually verify the increase in agitation of the water and the formation of the water vapour bubbles. The heat transferred to the water from an external source will raise its *temperature*.

If we sit in a hot room heated with a stream of vapour or simply hot air (let us give the examples of *batsu* or *sauna* in Scandinavia, although this habit is millenary and embedded in many cultures) we can feel the heat being transferred from the *surroundings* into our body (system), q > 0. But if then we choose to take a bath in a freezing pool or roll in the snow, we can feel the heat escaping from our body to the environment (q < 0).

9.4. The Zero Law of Thermodynamics

The capacity of transfer of heat from a body to another, gives origin to the *zero* law of thermodynamics. At this point one has to introduce the concept of *thermal equilibrium*. Let us consider two pieces of matter with different temperatures. If the two pieces are set in thermal contact, a flux of heat energy will surge from the object at higher temperature to the other, until they reach equilibrium. Of course, conditions of temporality, adequate contact, and proper isolation should be guaranteed.

For instance, we can put a cup of warm milk inside the refrigerator. Given sufficient time, the refrigerator thermally isolated, at some moment the cup of milk will have the same temperature as all other objects, including the air inside the refrigerator.

The zero law of thermodynamics is formulated as follows: if an object A is in thermal equilibrium with the object B, and if B is in thermal equilibrium with C, then A and C are in thermal equilibrium as well.

9.5. The Temperature Scales and When the World Went (Almost) Metric

Temperature scales have been created on the basis of the modification (linear, within certain limits) of some properties when heat is applied. Normally the chosen property is volume and gases, liquids and solids have been preferred depending on the range of temperatures aimed at. The main temperature scales are those of Celsius (Andreas Celsius, Sweden 1701–1744), Fahrenheit (Germany, Gabriel Daniel Fahrenheit 1686–1736) and Kelvin (named in honor of Lord William Thomson, 1st. Baron of Kelvin, Ireland, 1824–1907).

The Celsius scale defined as 0°C the freezing point of water and 100°C its boiling point. There are exactly 100 divisions in between. Yet, Fahrenheit gave a value of 32°F for the freezing point of water and 212°F for the boiling point. The interval is divided into 180 parts.

There are simple formulas to go from the temperature, T, marked on one scale, to the other:

$$T(°C) = \frac{5}{9}[T(°F) - 32] \qquad (9.2)$$

For scientific purposes, the absolute scale introduced by Lord Kelvin has become mandatory. This is the Kelvin scale. The connections between the Celsius scale and the Kelvin scale is

$$T(K) = T(°C) + 273.15 \qquad (9.3)$$

This means that water (at sea level) boils at 373.15 K. The value 0 K is the lowest temperature that might exist. *Almost* all movements are led to a stop. But in fact, the electron cloud of atoms and molecules is not at rest and even molecules have a residual vibration movement. This endless movement may seem odd, although consistent with quantum science.

There seems to be no higher limit for temperature values. Whatever hot star you can think of (millions of degrees), there is always something even hotter. Cosmologists have notions of unimaginably high temperatures at which the big bang is believed to have occurred. What are the properties of matter in such fantastic situations?

There is room to wonder, why, on the contrary, nature has set a lower limit to temperature, and, in fact, so close to our habitat as living beings. We know that high temperatures *destroy* while low temperatures, in many instances, *preserve*.

Temperature scales have in fact much to do with our technological civilization. By the middle of the past century, the world had (almost) "gone metric". In many countries, Fahrenheit went to Celsius, gallons went to litres, and miles went to kilometres. Not to forget that driving on the right of the road became mandatory in many countries (but less than one could imagine). Maybe this could be mentioned as one of the precursor and better aspects of globalisation. It was costly, however, culturally, financially and technically. Many machines, containers, maps, signs, tools and minds had to be changed.

9.6. Internal Energy and the First Law of Thermodynamics

The total energy of a system is denominated as *internal energy* (U). The internal energy is a state function in the sense that it depends on its properties (pressure, temperature, number of particles) but not on the manner the state was reached. U is an extensive property, meaning that it depends of the amount of substance. Thus, 2 g of some substance has twice as much energy as 1 g of the same substance in the same conditions.

If the system is isolated, its internal energy remains always constant; such is the first law of thermodynamics. Since the system is isolated, energy may be transformed from one form to another, but not destroyed or created. The first law has an empirical background and is not deduced from basic mathematical principles.

The internal energy U is the sum of the work (w) and heat (q) contributions:

$$U = q + w \qquad (9.4)$$

Since we shall lead with discrete (small and slow) increments in the variables, for convenience we shall rephrase Eq. (9.4) as

$$dU = dq + dw \qquad (9.5)$$

When the system is open, as is our body, we are constantly spending our internal energy through heat transfer and work. If we touch a cool object in winter, we transfer heat to it. When we go up a stair, we produce heat, transpire, and our bodies perform complicated biochemical work.[2]

These are examples of energy uses through work or heat transfer. We are deemed to recover this energy through food ingestion.

Some standards have been established on the necessities of food ingestion for different ages, types of work and physiological conditions. For this purpose the unit of *calories* is widely employed. Food samples are placed in devices called *calorimeters*, in which they are burnt and the amount of heat produced measured. The heat so produced is expressed in units called calories.

One calorie is the amount of heat necessary to raise the temperature of 1 g of water by 1°C. When we refer to the quantity of calories of a given food or drink we are implicitly counting the energy content of its chemical composition and structure.

Let us make a few comments on a subject very attractive to our fattening humanity.[3] The calorie content of the various food have been tabulated. For the principal varieties we have that:

- 1 g of fat produces 9 calories;
- 1 g of hydrocarbon produces 4 calories;

- 1 g of protein liberates 4 calories;
- 1 g of alcohol generates 7 calories.

In the specific case of the overweight, the "targets" of propaganda used to be mainly women (but nowadays, men and children are gaining more weight as well). Everybody may observe the significant interest in this matter by simply glancing at the magazines in the stands every week.

There are always a number of "new", unique, fantastic diets for losing weight at incredible rates. And you do not need to stop eating as much as you wish. There are endless lists of foods you should eat to be healthy and those you should avoid to prevent damage to health. The industry is always marketing new stuff.

9.7. The Second Law of Thermodynamics

The second law also has an empirical origin and provides us with guidance for the direction of events. Let us imagine we are viewing the following films.

- A closed room full of smoke and an open bottle on a table. At some moment, the smoke moves and enters the bottle.
- There is a ball on a soccer field; suddenly the ball lifts towards the sky.
- There is a cup of coffee and milk on the table. Unexpectedly, the coffee and milk separate in the cup.

In all the former cases we would be convinced that it was a joke and that the films were exhibited backwards. What we saw was far too unconvincing to be true: the ball suddenly flying up to the sky, the smoke like an obedient Genie entering the bottle and the milk choosing to gather in a corner of the cup.

But it is not always obvious in which direction things ought to occur and the second law will help us in the task of deciding in non-trivial cases. For this purpose another state function is defined: *entropy* (S). According to the second law a process can occur spontaneously if the entropy of the universe increases, that is in mathematical terms:

$$dS \text{ (system)} + dS \text{ (surrounding)} = dS \text{ (Universe)} > 0$$

In other words the entropy of the Universe will always grow. This statement does not imply that processes (or chemical reactions) with dS (system) < 0 cannot occur. They can occur, but they shall not be *spontaneous*. Thus we need further study to understand how we can induce our *system* to have dS (system) < 0, at the expense of dS (surroundings) increasing and, necessarily, dS (Universe) > 0.

Another way to understand this law is that not all the energy contained in our system can be transformed in work. Part will be transformed in heat and heat is disorder. And this heat will go to the surroundings and the universe entropy will grow.

Here is a very convenient manner for us to create a concept for entropy: degree of disorder. Indeed, we have intuitive and experimental notions of energy, work and heat. Let us create the notion that entropy is disorder and that in any spontaneous change disorder will grow.

In fact, we need to face another challenge, perhaps even more demanding than the second law. While in all processes the entropy of the universe does increase, in given systems it may decrease locally. And a non-minor case is life itself. Life involves a tremendous amount of order, organisation and stability. Since it started, whatever your own beliefs, matter adopted increasingly organised and ordered structures. Let us give as a privileged example: the human brain, which has billions of neurons with trillions of possible connections and unsurpassed memory storage capacity. Scientists have a good grasp of how many of these systems function and yet science has not even begun to understand consciousness. In the sense pronounced by Rene Descartes, mathematician and philosopher (France, 1596–1650): *cogito ergo sum* (I think, therefore I exist).

In particular, life is a local system in which entropy decreases, at the expense of the entropy increase of the surroundings and the whole universe. Whoever or whatever created man, probably did not anticipate to what extent this creature was capable of going beyond the second law in polluting and spoiling his habitat. Although, we think the reader may agree, also embellishing it.

The entropy variation of a system during a change may be determined, but the variation of the entropy of the surroundings is not amenable to measurement. Thus, deciding whether a process is spontaneous on the mere basis of entropy would be extremely difficult. To help us in this task, two more state functions are defined: *enthalpy* (H) and *Gibbs free energy* (G).

9.8. Enthalpy

Josiah Willard Gibbs (America, 1839–1903) introduced the "heat function at constant pressure" in 1875. Other scientists defined the modern term enthalpy later. The thermodynamic potential (enthalpy) was defined by the Dutch physicist Heike Kamerlingh Onnes (1853–1926) in the early 20th century in the following form:

$$H = U + pV \qquad (9.6)$$

Where U represents the internal energy of the system. Employing SI units, we have:

- H is the enthalpy (in Joules);
- U is the internal energy (in Joules);
- p is the pressure of the system (in Pascals, Pa), and
- V is the volume (in cubic meters).

Then H can be put to work to calculate the energetic balance of a chemical reaction and its preferred direction. This possibility of adding several steps of transformations to give a final result for the enthalpy is known as Hess' Law (Germain Henri Hess, Switzerland 1802–Russia 1850). The Hess' Law states that the state function and the difference between state functions are additive properties.

One example of application of Hess Law is the reaction between carbon and hydrogen to produce the gas propane, one of the components of kitchen gas.

$$3CO_{2(g)} + 4H_2O_{(l)} \leftrightarrow C_3H_{8(g)} + 5O_{2(g)} \qquad \Delta H° = 2220 \text{ kJ} \qquad (9.7)$$

$$3C_{(s)} + 3O_{2(g)} \leftrightarrow 3CO_{2(g)} \qquad \Delta H° = -1182 \text{ kJ} \qquad (9.8)$$

$$4H_{2(g)} + 2O_{2(g)} \leftrightarrow 4H_2O_{(l)} \qquad \Delta H° = -1144 \text{ kJ} \qquad (9.9)$$

The summation of the former three equations gives:

$$3C_{(s)} + 4H_{2(g)} \leftrightarrow C_3H_{8(l)} \qquad \Delta H° = -106 \text{ kJ} \qquad (9.10)$$

In the reactions above we employed the right-hand side sub-indexes $X_{(s)}$, $Y_{(l)}$, and $Z_{(g)}$ to indicate solid, liquid and gas phase, respectively.

Also, in the reactions (9.7) to (9.10) we have introduced a modification of nomenclature from H to H°. This implies that the enthalpy values are given for the standard conditions of temperature and pressure (STP), which guarantees us the possibility of useful and fair comparison between values. The STP accepted in chemistry are those of the International Union of Pure and Applied Chemistry (IUPAC), meaning that the enthalpy was measured at 0°C (273.15 K) and 100 kPa (this corresponds to 0.986 atm, which is the pressure at sea level). It may often be the case that the measurements were not realized at the STP, but then the necessary corrections are made so that comparisons may be valid.

Returning to equations (9.7) to (9.10) we can extract the following notions:

a) if $\Delta H° > 0$, the process absorbs energy in the form of heat; we call this type of reaction *endothermic*.

b) if $\Delta H° < 0$, the process is *exothermic*, it liberates energy to the surroundings in the form of heat.

Enthalpy is a composite energy term; it contains the thermal component (Entropy) and a component apt to perform work, called Gibbs free energy, as we signaled above. Thus we have

$$\Delta H = \Delta G + T \, \Delta S, \qquad (9.11)$$

where ΔH, ΔG and ΔS represent the variations in Enthalpy, free energy and Entropy, respectively. T is the absolute temperature.

9.9. Gibbs Free Energy

The Gibbs free energy provides us finally with a tool to predict when a process is spontaneous, at constant temperature and pressure. One may apply this criterion in various fields, like biochemistry, energy production, and synthesis.

The Gibbs free energy measures the maximum amount of work that can be extracted from a process, initiated in a state out of equilibrium and reaching a state in equilibrium. This is quantitatively provided by the value of ΔG, the force which commands the spontaneity of the reaction. In such manner, for the reaction

$$\text{Reagents} \leftrightarrow \text{Products}$$

ΔG (for the reaction) $= G_{products} - G_{reagents,}$ where $G_{reagents}$ is the Free Energy for the reagents, and $G_{products}$ is the Free Energy for the products.

For a process to be spontaneous we need $\Delta G < 0$, the process is then named *exergonic*. This means, it will perform work and dissipate energy to the surroundings. This is actually in accordance with the Second Law and any system out of equilibrium let on its own will tend to reach the equilibrium.

A non-spontaneous process occurs when $\Delta G > 0$, at the expense of energy provided from the surroundings. We call this process *endergonic*.

The standard free energy variation, $\Delta G°$, is defined for one mol of the reagent for T = 298K and p = 1 atm at pH = 0.

For applications in biology and physiology, it is common to change the standard form G° to G′, which is set at T = 298 K, p = 1 atm and pH = 7.0. The intention is to facilitate calculations and comparisons, since values of pH *ca.* 7.0 are common in such living systems.

Often we are interested in the value of ΔG at temperatures other than the standard. We may make the appropriate correction through the formula:

$$\Delta G = \Delta G° + RT \ln\frac{[\text{Reagents}]}{[\text{Products}]} \qquad (9.12)$$

with [Reagents] and [Products] representing the concentrations of reagents and products and R being the universal gas constant with the value 8.31 KJ/mol.

The ratio [Reagents]/[Products] = K is the equilibrium constant of the reaction. When K = 1 the tendency to form reagents or products is the same. Then the system is in equilibrium.

The ratio K allows for some interplay in inducing the reaction in one direction or another. For instance, if we arrange the system so that K > 1, the tendency of the reaction shall be the formation of products.

As we said before, living systems are open and never in equilibrium. They dissipate energy to the surroundings. Living organisms need free energy constantly to survive. Proteins are constantly destroyed and others have to be synthesised to replace them. DNA is continuously replicated. If the human body reached equilibrium, it would die.

In other words, metabolism is endergonic, thus it is not spontaneous. Still, these reactions do not occur in isolation, but associated to other, sufficiently exergonic reactions, so to make the former occur and life to be maintained.

When glucose is fabricated at the expense of solar energy, as we show in equation 9.1, other derivatives that save and carry energy are

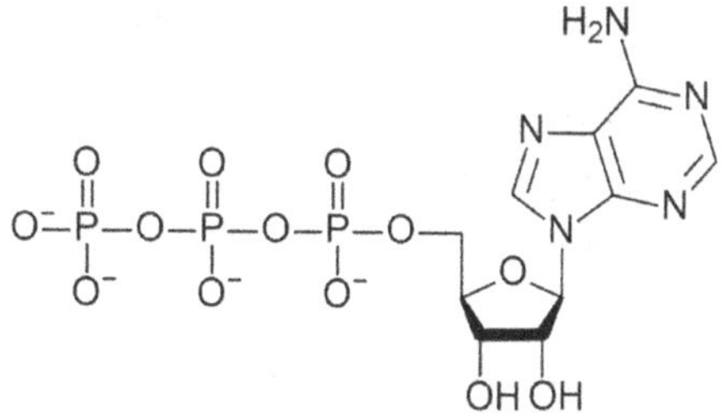

Fig. 9.1. The chemical structure of adenosine triphosphate (ATP).

also formed. A molecule which, is the great motor of almost all reactions that require energy is adenosine triphosphate (ATP, Fig. 9.1). The hydrolysis of ATP to adenosine biphosphate (ADP, the outer phosphate group is absent, as compared to ATP) provides the energy for numerous reactions.

For instance, the reactions described below are possible due to the hydrolysis of ATP to ADP. Using inorganic phosphate present, Glucose 6-phosphate is formed, at the expense of 3300 cal/mol:

$Glucose[C_6H_{12}O_6] + H_3PO_4$

$\rightarrow Glucose\ 6 - phosphate\ [C_6H_{13}O_9P] + H_2O \qquad \Delta G = 3300\frac{cal}{mol}$

The positive value of $\Delta G'$ suggests that the reaction will not proceed in the direction as signaled. Nonetheless, the simultaneous hydrolysis of ATP to ADP makes the occurrence of both reactions possible i.e.,

$C_6H_{12}O_6 + H_3PO_4 \rightarrow C_6H_{13}O_9P + H_2O \qquad\qquad \Delta G' = 3,3\frac{kcal}{mol}$

$C_{10}H_{16}N_5O_{13}P_3[ATP] + H_2O \rightarrow C_{10}H_{15}N_5O_{10}P_2[ADP] + H_3PO_4 \quad \Delta G' = -7,3\frac{kcal}{mol}$

$C_6H_{12}O_6 + ATP \rightarrow C_6H_{13}O_9P + ADP \qquad\qquad\qquad \Delta G' = -4000\frac{cal}{mol}$

Thus, adding the first two equations we realize that the reaction does occur with the liberation of -4000 cal/mol.

9.10. The Third Law of Thermodynamics

The third law of thermodynamics states that at a temperature of zero (absolute) the entropy of a perfect crystal is zero. At 0 K all movement other than that required by the uncertainty principle of quantum mechanics will cease.

The closest structure we know of as a perfect crystal is diamond, a specific arrangement of carbon atoms.

The entropy of diamond provides us with the origin of the entropy scale for elements and compounds.

Graphite is another allotropic form for carbon (allotropes are different forms that an element may adopt: for instance, sulphur is an element with a variety of allotropes: S_2, S_3, S_8, etc.). It so happens that the transformation

$$C \text{ (diamond)} \rightarrow C \text{ (graphite)} \tag{9.13}$$

has a Gibbs free energy lower than zero; thus, reaction (9.13) is a spontaneous reaction. Nevertheless, it is a very slow reaction, it is hard to think of slower examples.

So if you gain a diamond necklace and you decide to guard it safely, you may be sure that your great-great-grand-daughters will be able to use it much before it becomes graphite. Contrarily, some reactions are so fast that they occur before we can perceive. Here we encounter another aspect of chemical reactions: they proceed with variable velocity and one of the tasks of chemistry is to be able to control or at least influence the velocity.

9.11. Chemical Kinetics

Chemical kinetics is the study of the rate at which chemical reactions occur and the various factors which may influence such a rate.

In our daily life reactions may be slow, fast, and moderate. At times they occur instantaneously, as in an explosion. At various instances we manipulate the speed of reactions. When we conserve food in the refrigerator we seek to retard the decomposition of the organic matter. When we use the pressure cooker, we wish to accelerate the cooking.

In order to be able to control a chemical reaction we need to learn what factors govern it.

9.12. The Chemical Reaction

For a chemical reaction to occur there needs to be contact between the reagents in the first place and they ought to have what is called chemical affinity. Not every collision will result in a reaction; it is necessary that the collision be effective, with adequate orientation and sufficient energy to break the formerly existing linkages and generate new chemical bonds.

The smaller amount of energy necessary for a reaction to occur is named *activation energy*. This energy is necessary, independent of the reaction being exothermic or endothermic. The lower the activation energy necessary, the faster will the reaction occur.

When the molecules collide with the proper orientation and with the necessary energy, an instable intermediary will be formed; we call it the *activated complex*. In the activated complex, the linkages with origin in the reagents are weakened, while new bonds start to be formed.

Figure 9.2. illustrates the former discussion for both an exothermic and an endothermic reaction.

9.13. The Velocity of a Chemical Reaction

Let us use for our discussion a reaction in which all reagents and products are in the gas phase. We may represent the reaction as

$$aA_{(g)} + bB_{(g)} \rightarrow xX_{(g)} + yY_{(g)} \tag{9.14}$$

In the reaction above A and B are the reagent molecules and a and b the number of each such molecule entering the reaction. X and Y are the produced molecules and x and y the respective number of them in the reaction.

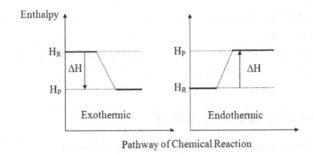

Fig. 9.2. Schematic representation for the pathway of chemical reactions: the exothermic case and the endothermic case. The symbols H_R and H_P stand for the enthalpies of reagents and products.

A real reaction example will facilitate the understanding of Eq. (9.14). Let it be the reaction of combustion of methane to produce carbon dioxide and water:

$$CH_{4(g)} + 2O_{2(g)} \rightarrow CO_{2(g)} + 2H_2O_{(g)} \qquad (9.15)$$

Then we have that CH_4 is A, O_2 is B, CO_2 is X and H_2O is Y and a = 1, b = 2, x = 1 and y = 2.

The reaction will proceed as in Fig. 9.3, with the proportion of products increasing with time, while the reagents diminish simultaneously.

The formal mathematical expression for the velocity, v, of the reaction (9.14) is given by

$$v = k \, [A]^{\alpha}[B]^{\beta} \qquad (9.16)$$

where k is the rate constant characteristic for each reaction (needs to be determined experimentally) and α and β are reaction orders; these reaction orders also need to be experimentally established.

The brackets for A and B, i.e., [A], [B], are symbols for the concentration for the respective compounds.

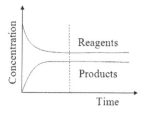

Fig. 9.3. Variation of the concentration of reagents and products with time for a chemical reaction.

9.14. Factors Which Influence the Reaction Velocity

From the expression for the velocity of a chemical reaction, *v*, as shown in Eq. (9.14), we can influence such velocity only through the concentration of the reactants or the value of the rate constant *k* itself.

Svante Arrhenius (Svante August Arrhenius, Sweden 1859–1927) obtained a mathematical expression for *k* which put in evidence the influence of the temperature and other factors for the rate constant. The formula deduced by Arrhenius was

$$k = Ae^{-Ea/RT} \qquad (9.17)$$

In Eq. (9.17) Ea is the activation energy (see Fig. 9.4), R is the gas constant, and *T* is the absolute temperature.

The factor A is the Arrhenius parameter and encompasses information on both the collision frequency and orientation of the systems.

One major task of chemistry is the synthesis of new products (fertilisers, plastics, solvents, paints, medicines). Our daily life is impregnated with the products of chemical industry. It is certainly desirable that the synthetic paths are simple, non-pollutant and not expensive. In this context, the faster the reaction, the more efficient is the procedure. We comment below on some of the many factors which may accelerate a reaction.

Temperature. The consideration of Eq. (9.17) indicates that when the value of T increases, the value of k also increases. The physical meaning of it is that the number of collisions will increase and the increase of kinetic energy will make more molecules reach the minimum energy to react, Ea.

Pressure. When we work in gaseous phase the increase of pressure will have the same effect as the increase of concentration (equation (9.16)); the rate of collisions between the molecules shall rise.

Concentration. In a similar manner, and also obeying equation (9.16), higher concentration will favor faster reaction.

Light. Some reactions occur faster under illumination or even become possible only with light. In fact, such reactions form a vast and rich branch of chemistry denominated *photochemistry*. For instance, a mixture of H_2 and Cl_2 does not react in the dark, but might explode when exposed directly to solar light:

$$H_{2(g)} + Cl_{2(g)} + light \rightarrow 2HCl_{(g)} \qquad (9.18)$$

A particular and most crucial photochemical process is photosynthesis, to which we refer at the beginning of this chapter and in Sec. 10.6 of the next chapter.

It cannot be emphasised enough that life on earth would not exist in the form we know it without photosynthesis.

Catalysts. The use of *catalysts* allows the chemist and nature to accomplish structures that would seem elusive if the laws of thermodynamics and kinetics would have absolute control. Catalysts are atoms or molecules in gas, liquid or solid state which allow sort of a bypass or shortcut for a chemical reaction.

To explain how the catalysts work, let us refer to Fig. 9.4, which illustrates the role of the activated complex for the reaction and the need to provide the necessary energy to surpass the activation energy, Ea. The function of the catalysts is to create a new route for the reaction, in which the activation energy, Ea' has a lower value than it has for the normal reaction. In Fig. 9.4 we show how a reaction with and without a catalyst proceeds.

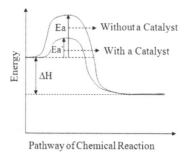

Pathway of Chemical Reaction

Fig. 9.4. Reaction pattern for a reaction with a catalyst and a reaction under common conditions. Ea' and Ea are the respective activation energies. ΔH is the enthalpy variation for this exothermic reaction example.

In the chemicals industry, catalysts, sometimes metal surfaces simplify and save time and investment for the fabrication of products. In principle, the catalyst emerges as it was before the reaction, unaffected and able to be used over and over again. In practice, its properties weaken with repetitive use and the catalyst has to be replaced or recovered. We know well that the catalysts for better combustion in the car exhaust need periodic maintenance or replacement.

In the case of living beings, the role of catalysts is played by the enzymes. In Chapter 11, we focus on the structure of proteins and the mechanism of action of enzymes, more specifically in Sec. 11.3.

Inhibitors. These are substances that decrease the velocity of reactions by augmenting the activation energy. Thus, they play the opposite role of catalysts.

In fact here we are talking about a vast field of preservatives, or, more generally, additives, in food, pharmaceuticals, beverages and perishable products in general. There are many preservatives for food, not only to retard decomposition, but to prevent growth of fungi, bacteria, and insects. The subject is complex and controversial, with many health concerns. Additives may also have other purposes as dyes or flavour enhancers. Different countries have different legislations for additives.

Physical state of the reagents. In general terms, gases react faster and more easily than liquids and the last faster than the solids.

Electricity. Electrical energy may be used to increase the kinetic energy of the molecules and in such a way increase the velocity of the reaction.

A spark initiates the ignition of gasoline in automobile engines and can also initiate the reaction between hydrogen and oxygen gases:

$$2H_{2(g)} + O_{2(g)} + \text{electric flash} \rightarrow 2H_2O_{(g)} \qquad (9.19)$$

The spark provides sufficient energy for a few H_2 and O_2 molecules to surpass the activation energy. Since the reaction itself generates a large amount of energy, this is sufficient to induce the reaction in the bulk of the gas mixture.

CHAPTER 10

Chemistry and Energy Sources

10.1. Preliminary Considerations

The energy issue in our world is one of tremendous concern and endeavour. Mankind employs it, wastes it or envies. We are bombarded by this drama in the media. Enormous catastrophes occur (wars, broken dams and pollution). The reader will, no doubt, have his/her own opinions on this matter. We shall not take here any partisan or radical position for the tremendous geopolitical challenge of humanity and the planet: energy, water, food, minerals and other, not to mention that governments, scientist, economists, business people are far from any consensus. We will rather focus on chemical sources of energy and mention other alternatives. Anyway, we shall come back to this issue in section 10.8.

A major concern nowadays for many is to produce energy as cleanly as possible.

Petrol, gasoline, often called simply gas, natural gas, coal, ethanol, wood and other combustion materials still dominate industry, transport, heating or cooling, with petrol being the most important nowadays. The heat produced by combustion is sometimes employed directly (as in a gas stove). Or transformed into mechanical work (as in the old vapour machines, as a nostalgic example) or then used to generate electricity as in the thermoelectric plants. Electricity can be transported although transport of electricity is expensive and there is a big loss over long distances. Neither have we an efficient way to store it, yet.

Combustion is a chemical process which we shall address below. The hydrogen fuel cells seem a cleaner chemical perspective and we shall describe this option later. Before describing combustion, we mention other forms of generating energy, the list not being necessarily complete.

10.2. Solar Energy

Man-made devices for the technological application of solar energy are at the beginning of discovery and implementation. One may speculate that all our present concerns on energy sources could be one day resolved through this enormous source of energy. Every minute, 2.0 calories/cm^2 originating from the sun reach the outer earth atmosphere. There is plenty of nuclear fuel in the sun, until it eventually becomes a white dwarf, that is.

Our planet, as it is, and life itself, exists merely because we receive the heat and light from the sun. Climate, day and night, and life are all due to solar energy. Almost all forms of life have their ultimate origin in photosynthesis.[1]

Batteries of vehicles, heating and illumination of interior habitats, machines and so on, could be activated through solar produced energy. One does not need to think of large and expensive solar cells. According to a recent prognosis by IBM scientists, thin conducting polymer films could cover streets, walls and windows for the generation of electrical energy from solar source.

10.3. Mechanical Energy Production

When traveling by road in some places one sees many windmills with their gently rotating sails. The mills' arms are moved by the wind (this is *Aeolian* energy) and this mechanical energy is processed into electricity.

Earth's gravitational attraction on water is presently the second source of energy after combustion. It is supposed to be a clean way to produce electricity through the many small, medium and giant hydroelectric plants in many countries.

Actually, the forced migration of human populations, and the mortality of flora and fauna for the construction of dams, are important humanitarian and ecological impacts produced as a consequence of this form of energy generation.

A further application of gravity is the attraction between the earth and the moon, which produces sea tides; this motion is in the end transformed into electrical current.

10.4. Nuclear Energy

Nuclear power plants employ mainly nuclear fission of uranium 235, $^{235}U_{92}$, as fuel. In the first place, the uranium has to be extracted (or bought) and transported. The natural composition of uranium has almost 99.3% of the 238 isotope, 0.7% of ^{235}U and a very small amount of the 234 isotope. Through various chemical or physical processes the uranium has to be "enriched", that is, the proportion of the $^{235}U_{92}$ isotope has to be raised in the mixture to the level required by the nuclear reaction. The reaction is initiated by bombardment of the enriched uranium with neutrons:

$$^{235}U_{92} + n \rightarrow {}^{236}U_{92} \tag{10.1}$$

The new nucleus absorbs the neutron, but soon splits into two new atomic species:

$$^{236}U_{92} \rightarrow {}^{144}Ba_{56} + {}^{89}Kr_{36} + 3n + 177 Mev \tag{10.2}$$

Other than the energy produced by the former reaction, three new neutrons are expelled, initiating the chain reactions, which in turn will split other uranium-235 atoms.

Other fission pathways also occur. The radioactive waste has to be discarded in such a way as to prevent contact with humans, animals, nature in general. The most dangerous fission fragments produced are Cesium-137 and Strontium-90. Both are radioactive with a mean life of some 30 years. To make things worse, Cs may be absorbed as if it were potassium by living beings (see the alkaline family in the periodic table).

Accidents or attacks in nuclear plants may lead to disasters. The biggest accidents seem to have been that of Chernobyl in 1986, in the former Soviet Union and, with less lethal consequences, in Three Mile Island in the USA in 1979. In spite of these drawbacks, an increasing number of countries have been building nuclear energy plants, although we may say the rate is slowing down. At the present rate, nuclear energy is not going to surpass fossil fuels or hydroelectric plants. It is claimed that waste disposal and security are being satisfactorily handled. But, as a matter of fact, the final destination of the nuclear waste has not been solved by any country. As there are other alternatives discussed in the present chapter, this is still an open issue.

10.5. Chemical Energy: Combustion

The heavy hydrocarbons of petrol (often called oil) are broken (cracked) in several hydrocarbons of variable molecular weight and the fractions separated by distillation. The various fractions so obtained are employed separately.

Since the beginning of the twentieth century, transport has relied on liquid fuel, as electrical tramways and vapour locomotives were displaced. This is true for air, water and surface transport. In the particular case of automobiles, the combustible is gasoline. In the USA it is common to see "gas stations", although what they sell is gasoline. Incidentally, in Brazil there is a fleet of cars moved really on gas (methane, CH_4, and other light hydrocarbons), which is, at some moments of price oscillations, cheaper than alcohol, which is cheaper than gasoline.

One of the components of gasoline is the hydrocarbon octane, C_8H_{18} with several *isomers*. One of the isomers of octane is the linear chain called n-octane, with the formula

$$CH_3 - (CH_2)_6 - CH_3 \qquad (10.3)$$

The complete combustion of octane is given by the reaction

$$C_8H_{18} + 12.5O_2 \rightarrow 8CO_2 + 9H_2O \qquad (10.4)$$

This reaction produces 33.76 kJ for each mol of octane. One mol of octane is equal to some 112 g; considering the density of n-octane, 0.70 g/cm^3 or 0.70 g/mL, we have consumed 0.16 L (less than half a gallon) of gasoline. Using very rough average values, we have burnt some 30 USA cents to cover 1 km (half a mile, say).

As we signaled in Chapter 4, in fact the best combustion is obtained with the branched octane molecule, i.e., iso-octane. The quality of gasoline is usually expressed as the proportion of iso-octane present. Other components of gasoline would have a tendency to burn incompletely.

The incomplete combustion leads to the production of the gas carbon monoxide (CO), as in the reaction

$$CH_4 + 3/2O_2 \rightarrow CO + 2H_2O \qquad (10.5)$$

Carbon monoxide is a colorless and odorless gas which when inhaled, binds to haemoglobin in the lungs, blocking the site which would be normally occupied by oxygen, thus producing asphyxia; we show the main biochemical aspects of respiration in Chapter 11.

At one time, it was discovered that the tetraethyl lead molecule, $Pb(C_2H_5)_4$, added to gasoline would produce the same benefit as a height proportion of iso-octane, i.e., it is an anti-detonant. Nonetheless, the presence of lead in the exhaust gases is toxic and, after some decades, the use of lead compounds is being abandoned.

In the last ten years or so, the tendency for gasoline propelled cars has been to insert a catalyst at the end of the exhaust system to complete the combustion.

To make things worse, the element sulphur is present in petrol. If a purification process is performed, sulphur content in gasoline can be decreased (resulting in the residual elemental sulphur becoming a problem in waste disposal). Otherwise, sulphur will burn together with the petrol and enter the atmosphere as sulphur dioxide, SO_2, and sulphur trioxide, SO_3, gases.

The above-mentioned sulphur gases are toxic. With water in the air they will form sulphurous acid, H_2SO_3, and sulphuric acid, H_2SO_4. Gases of nitrogen, also present (NO, NO_2, N_2O_5) as motor combustion products, will form nitric, HNO_3, and nitrous, HNO_2 acids. These chemicals form *acid rain*, corroding our lungs and monuments.

10.6. Bio Fuels: Alternative Combustion

Although all we described above seems bad enough, still the big issue today is carbon dioxide, CO_2, a necessary product of combustion as shown in equation (4.10). In 2006 Japan produced 11.5×10^6 vehicles, the USA 11.3, China 7.2, and so on, totalling some 10^7 vehicles. Certainly indeed, many other sources produce CO_2.

The accumulation of CO_2 in the atmosphere leads to the crucial issue of global warming. This is a most challenging crossroad that humanity faces. Enormous scientific, geopolitical and economic matters and antagonisms are at stake. There is no predictable outcome between many conflicting interests and dogmas. Even the complete discussion of the strictly technical alternatives and implications is not unanimous. Still, we shall address routes that could possibly retard CO_2 production, if that is possible.

One major route is the use of "renewable" sources of energy (in contrast to *fossil* originated products, such as petrol or coal, which ought to dry up eventually, if present day rate of consumption continues). From the global warming point of view, what matters is that these sources *consume* CO_2 for its production (specifically, for the plant to grow). So far, the chosen fuel of vegetable origin has been ethyl alcohol or ethanol, C_2H_5OH, with formula

$$CH_3-CH_2OH.$$

The combustion of alcohol proceeds as

$$C_2H_5OH + 3O_2 \rightarrow 2CO_2 + 3H_2O \tag{10.6}$$

The plant most abundantly grown for obtaining sugar is sugarcane. Ethanol is produced after fermentation of sugar and distillation so as to produce pure alcohol. Two countries are presently producing large amounts of alcohol: Brazil, employing sugar-cane as a source and the USA, using corn. In the case of corn, a side effect is the rise in the price of corn and heavy government subsidies are necessary for this crop. Also, in both cases, the use of agrochemicals or rudimentary agricultural techniques is to be lamented. Still, some decrease in the dependency on petrol is undeniable. More important yet, the CO_2 balance is clearly favourable to alcohol as compared to petrol since the plants "pay the fee" in advance, consuming CO_2 for their growth.

Automobiles may run exclusively on alcohol, as is the case of a large proportion of cars in Brazil; this is certainly significant if we consider that only the city of São Paulo has a fleet of 10^6 cars. Nowadays, cars are produced so as to be able to function with any fuel: the flex-fuel models. Nevertheless, presently gas stations across the world will provide only gasoline or diesel. But in the future, they could provide also alcohol, or hydrogen, or plugs for charging the car batteries, or, even water for the fuel cells.

The use of ethanol in some countries other than Brazil has been so far limited to the employment of *gasohol*, gasoline with a variable (may be up to 25%) proportion of alcohol added. Another current alternative under scrutiny is *biodiesel*, diesel with some proportion of vegetable originated oil.

There are various aspects to be considered regarding the use of alcohol instead of gasoline. But in this section we aim at the carbon dioxide balance in the atmosphere. The growth of sugar-cane, corn or other plants consumes CO_2 through photosynthesis. Thus, in such a way, at least part of the amount of the CO_2 gas which will be produced by combustion is consumed beforehand.

Green plants synthesise glucose ($C_6H_{12}O_6$) with the aid of chlorophyll and solar radiation as shown in the equation (10.7) below

$$6CO_2 + 6H_2O + \text{solar energy} \rightarrow C_6H_{12}O_6 + 6O_2 \qquad (10.7)$$

Thus, green plants (and cyanobacteriae and algae too), when growing, consume water and carbon dioxide to produce plant building molecules, such as sugar, cellulose, lignin, and release oxygen gas to the atmosphere.

10.7. Fuel Cells

A hydrogen fuel cell (see Fig. 10.1) will have H_2 gas injected at the anode and oxygen from the air entering the cathode. The half-reactions will be:

$$\text{Anode: } H_2 \rightarrow 2H^+ + 2e^- \qquad (10.8)$$

$$\text{Cathode: } \tfrac{1}{2} O_2 + 2H^+ + 2e^- \rightarrow H_2O \qquad (10.9)$$

And the overall reaction:

$$H_2 + \tfrac{1}{2} O_2 \rightarrow 2H_2O \qquad (10.10)$$

The electrons flow from the anode to the cathode, producing eventually electrical work. In the cathode the electrons reduce oxygen and combine with hydrogen ions to form water, which is expelled through the exhaust. The reaction of oxygen with hydrogen to produce water occurs with the aid of catalytic materials such as platinum. The protons are transported through an appropriate medium (porous, polymer, electrolyte) from the anode to reach the cathode.

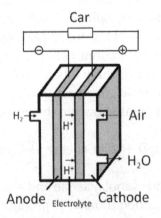

Fig. 10.1.   Scheme of a fuel cell showing the flux of chemicals and electricity.

The former mechanism would suggest that we need a source of hydrogen supply in the cell. Hydrogen storage rechargeable materials have been indeed produced, as is the case of metal hydrates. In these compounds, the metallic positive ion, say Na^+ forms a salt with the negatively charged hydrogen atom, H^- and the resulting compound is NaH, sodium hydrate. A salt which has been shown to be appropriate for the storage of hydrogen is the hybrid hydrate $NaAlH4$; aluminum is present as the Al^{+3} ion, so the whole molecule is neutral. This salt releases hydrogen under soft heating and can be regenerated for new use. These materials would play the role of "batteries" in hydrogen-propelled vehicles. In several cities across the world buses are using such batteries for hydrogen cells in electric engines.

But then why not generate hydrogen in the system itself. One could use water and through *electrolysis* separate the water into oxygen and hydrogen. The electrolysis process is produced when an electrical current is passed through two electrodes immersed in the water. The effect is to split H_2O in the component gases H_2 and O_2 in a process opposite to the fuel cell, as shown in equation (10.11) below. The hydrolysis reaction is

$$H_2O \rightarrow H_2 + \tfrac{1}{2}O_2 \qquad (10.11)$$

The hydrogen gas produced may be employed to feed the fuel cell again. It may appear as sort of self-defeating to produce water in the fuel cell while generating an electrical current to produce work, and then to use electricity to split water into hydrogen and oxygen again. But intelligent combinations with solar cells or other sources of electricity are certainly possible and may be economically viable.

Solar cells could be employed to produce hydrolysis, or, alternatively, solar energy could be brought to generate hydrogen through processes similar to photosynthesis. For instance, some species of algae could be induced to deviate from their usual photosynthetic path to accumulate hydrogen.

Another chemical possibility under scrutiny is the catalytic conversion of alcohols as methanol or ethanol to produce hydrogen. For instance, in the case of ethanol, the general reaction is

$$C_2H_5OH + 3H_2O \rightarrow 6H_2 + 2CO_2 \qquad (10.12)$$

Thus in this case the generation of hydrogen is not as clean as it is when hydrogen is obtained from water.

Many cities have buses powered by hydrogen electric engines, or direct electricity circuits, making the old well known trolley buses gain a new ecological support. The nostalgic tramways using electricity may very well cease to be a thing of the past.

10.8. Is Global Warming Real?

As we asserted in the introduction of this chapter, there is no hint of any consensus in energy and pollution matters. The alert reader perhaps knows that even the issue of global warming itself is being contested by some politicians and scientists.

A number of respected scientists doubt that global warming is a man-made effect. They argue that the increase in CO_2 in the atmosphere does not parallel temperature rise and that a major solar activity may explain the warming.

As chemists and authors of this book, we do not consider it fair to open the question completely for the non-specialised reader to pick his or her side. First, the reader may have noticed that all along the chapter we advocate for pollution control. It is also important to remember that almost all major scientific societies across the world are alert to the global warming risk.

On the other hand, very little is done globally. The almost endless meetings of the 8, the 20, the UN, etc., are only empty words leading to nothing.

Few countries and cities are taking serious steps. The real issue is not whether global warming exists but whether it is man made or not. But disaster may be just around the corner. There is visible evidence: pollution with plastics and oil spills in the oceans, cities where poison is breathed instead of air, cities which have no place to put their garbage, dead fish in polluted rivers, birds impregnated with oil and dying, not forgetting humans with neither food nor drinking water due to changes in the global climate.[2]

We do not yet know of any real solutions for pollution. We merely bet on substitution of one pollutant for another, hoping that the next is less pollutant than the last. We may substitute gasoline for ethanol; but growing sugar cane or corn needs fertilizers. You may use fuel cells, but you need catalyzers as platinum. You may use batteries to move cars but you have to dispose or the lithium. Solar cells will need niobium.

Whatever solution you might think of, you will find a drawback at some point in the road.

CHAPTER 11

Chemistry and Living Beings

11.1. Preliminary Considerations

As in previous chapters, some interesting subjects of a very vast area of chemistry shall be selected and explained in some detail. The purpose is to reveal part of the marvelous world where life and chemistry meet. The preferences and experience of the authors have influenced the chosen subjects.

Much of the material to be explained may be classified under the denomination of *biochemistry*, that is, chemistry of living beings. Nevertheless, physics is also well at the basis of processes in living beings and some of the notions given below in sections 11.4. and 11.5. may be classified as *biophysics*.

Let us begin by facing the hopeless task of understanding from the scientific standpoint what life itself actually is. An intuitive common sense notion of life, every one of us has. But an unequivocal definition is elusive. As a matter of fact, is it necessary or even possible to *define* life?

Attempts to define life are rather lists of characteristics which organised entities should exhibit to be considered alive. But, so far such lists leave always some loose bricks. For instance, one property should be the ability to reproduce; still, worker ants are sterile but we believe they are alive indeed (and very busy!). On the other hand, viruses can reproduce (or replicate) but they do not grow.

This subject is captivating. Can computational patterns simulate life or realise life? Can we create a cell in the laboratory?[1] At what moment does life start: when a male sperm cell enters a female ovum and the embryo is formed? This last issue goes well beyond science and concerns human choices.

Maybe we can simply agree that we believe that the human being is the most complex form of life known so far to us.

Let us list some parts of the human body with different chemical composition, anatomic structure and functions: muscles, bones, nails, hair, teeth, blood, skin, kidney, liver, lungs, pancreas, bladder, and many others. We close the list with what we believe to be the most sophisticated structure we have: the human brain.

The former structures are composed of a large number of chemical compounds, some very complex: proteins, lipids, hydrocarbons, calcium phosphate, etc. And, of course, water; we are made of 45–70% water in weight (depending on sex, age, and body mass).

11.2. Proteins

Proteins are macromolecules (meaning large molecules) or we can also consider them as polymers, that is, built as chains of many units; these units are called *amino acids*. Since there are 20 different molecules in the amino acids family and proteins are built by an enormous arrangement of sequences of them, we may consider proteins as *aperiodic* polymers, with the prefix of Greek origin: *a* implying negation or lack of periodicity.[2]

Proteins are omnipresent in living beings, constituting the basic material of muscles, enzymes, cell membranes, transport molecules in blood and many other structures and functions.

The basic structure of an amino acid is shown in Fig. 11.1. In Fig. 11.2 below we show the formula of some common amino acids: (a) the simplest of them, glycine, in which $-R = -H$; (b) alanine, $-R = -CH_3$, (c) lysine, with two amino groups, (d) cysteine, which contains a sulphur atom, $-R = -CH_2-SH$; and (e) phenylalanine (see Fig. 11.2).

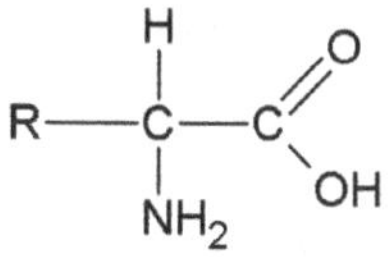

Fig. 11.1. The basic structure of an amino acid is illustrated. R stands for various possible substitutions which may be aliphatic, cyclic or hetero-atomic.

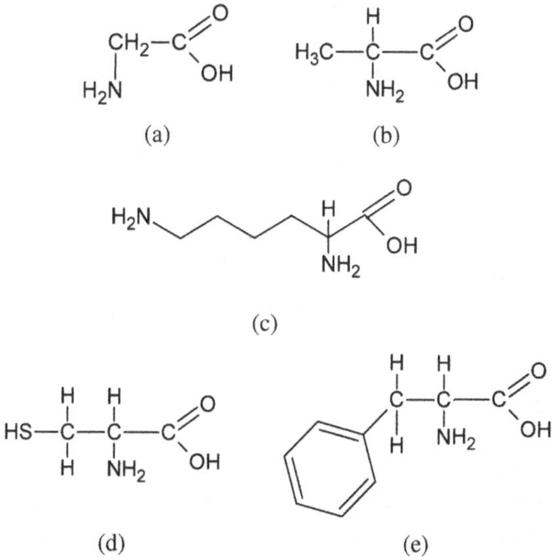

Fig. 11.2. The chemical structure of five amino acids: (a) glycine; (b) alanine; (c) lysine; (d) cysteine; and (e) phenylalanine.

The long chains of proteins are formed by the reaction of the carboxylic group (–COOH) of one amino acid with the amino group (–NH$_2$) of another amino acid, as shown in the scheme below:

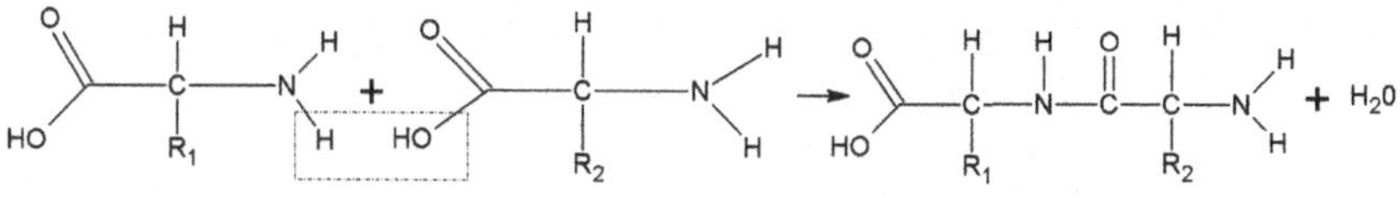

The size of a protein chain is very variable. It may merely have a few hundred amino acid groups or (the shorter units are also called

peptides) up to 30,000 amino acid units. The sequence of amino acids in a given protein has been given the name of a primary structure, being different for each particular protein. These chains of amino acids arrange helicoidally giving place to what is called the secondary structure of the protein. Further, these helicoidal arrangements fold in various manners, giving lieu to what is called the tertiary structure of a protein. What is called quaternary structure of a protein, involves in fact the association of two or more protein chains.

Such richness of structures is one of the bases of the multiple functions of proteins in the living cell. To start, one of the materials for the construction of the cell is protein; further we have the muscular contraction, the structure of receptors for hormones and drugs, the transport of materials in the blood, the control of genes and many others.

For various functions, proteins shift between several structures, that is, they undergo conformational changes, returning usually to the original form in a reversible way, after the task is completed.

11.3. Lipids

Lipids are a very wide and diverse class of molecules of biological origin that have in common low solubility in water and high solubility in non-polar organic solvents; some common organic solvents are listed in Table 11.1.

Table 11.1. Common Organic Solvents

Solvent	Compact formula
Acetic acid	$C_2H_4O_2$
Acetone	C_3H_6O
Benzene	C_6H_6
Carbon tetrachloride	CCl_4
Chloroform	$CHCl_3$
Cyclohexane	C_6H_{12}
Ethanol	C_2H_6O
Hexane	C_6H_{14}
Methanol	CH_4O
Toluene	C_7H_8

We may define a solvent as a substance which is liquid at room temperature (usually this means 20°C) and in which we can dissolve a solid, another liquid and eventually a gas. In our habitat, water is a solvent by excellence. We well know how easily it can dissolve sodium chloride, sugar or ethanol. But it will not dissolve a drop of oil. Oil is a lipid.

Water is a polar solvent, meaning that the water molecule has a negatively charged region in the vicinity of the oxygen atom and a positively charged region where the hydrogen atoms are. Some of the solvents listed in Table 11.1 are in fact polar, in the same sense as water; this is the case, for instance, for ethanol (CH_3-CH_2OH) or for acetic acid (CH_3-COOH). On the other hand, in the list of Table 11.1, nonpolar solvents, i.e., with homogeneous electric charge distribution are molecules such as benzene, cyclohexane, and carbon tetrachloride.

The very broad definition of lipids as given at the beginning of section 11.3. leaves room for an enormous variety of compounds under this classification. We shall address in this section specifically the following groups of chemicals: fatty acids, triglycerides and phospholipids.

Two features of lipids are commonly pointed out as their main functions in living beings. First, they serve as source or storage of energy. Their controlled "combustion" (see Chapter 10, section 10.5) in the cell produces heat, leaving water and carbon dioxide as residues.

The other function of lipids is structural, as these substances, together with proteins, are the basic molecules for the construction of the cell membranes.

When we discuss the mechanism of vision in vertebrates (Section 11.6) we shall refer to the molecule of retinal, which is also a lipid.

11.3.1. *Fatty acids*

A fatty acid is composed of a hydrocarbon chain (usually 12 to 24 carbon atoms) terminated by a carboxyl group ($-COOH$) as in the acetic acid, CH_3-COOH. Very common members of the family are stearic acid, $CH_3(CH_2)_{16}COOH$, and palmitic acid, $CH_3(CH_2)_{14}COOH$. Stearic and palmitic acids are examples of saturated fatty acids, that is, there are no double bonds in its hydrocarbon chains.

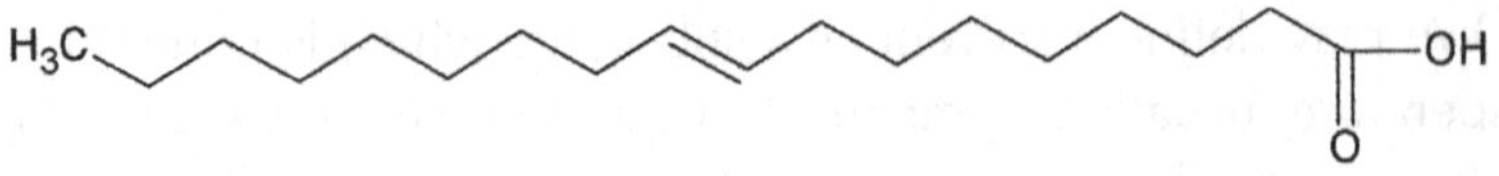

(a)

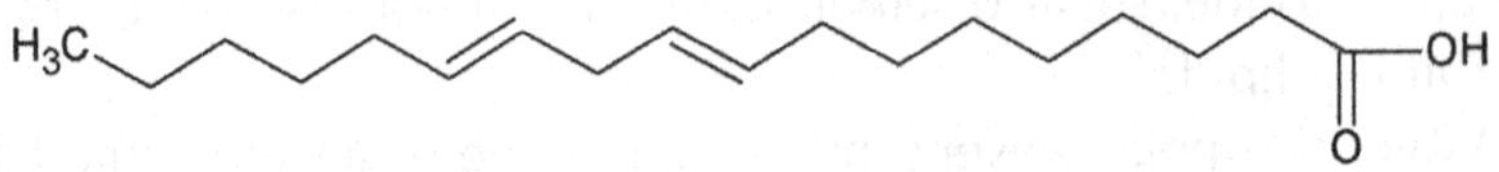

(b)

Fig. 11.3.   The structural formulas of (a) oleic acid and (b) linoleic acid.

In nature one also finds unsaturated fatty acids, the most common being oleic acid (see Fig. 11.3a). It may also occur that more than one double bond is present in the chain, as in the case of linoleic acid (see Fig. 11.3b).

11.3.2. *Triglycerides*

The large amounts of fatty acids encountered in biological systems are rarely found as free acids, but rather esterified to glycerol a tri-hydroxy alcohol to form triglycerides or triglycerols.

Esterification is a reaction between an organic acid and an alcohol to form a salt, an ester in the organic case, as shown in the scheme below:

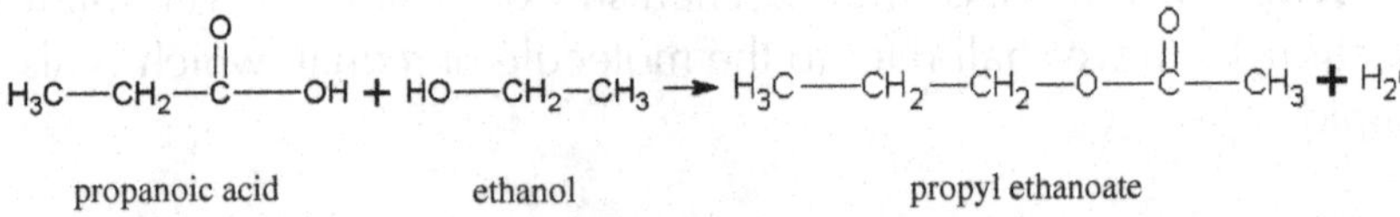

propanoic acid   ethanol   propyl ethanoate

In nature the esterification of fatty acids occurs mainly with the alcohol glycerol (Fig. 11.4a). Glycerol combines with three fatty acid units (often of the same kind) to form triglycerols, the energy storage in biological systems. Figure 11.4b shows the chemical formula of tristearin, in which the fatty acid forming the ester is stearic acid.

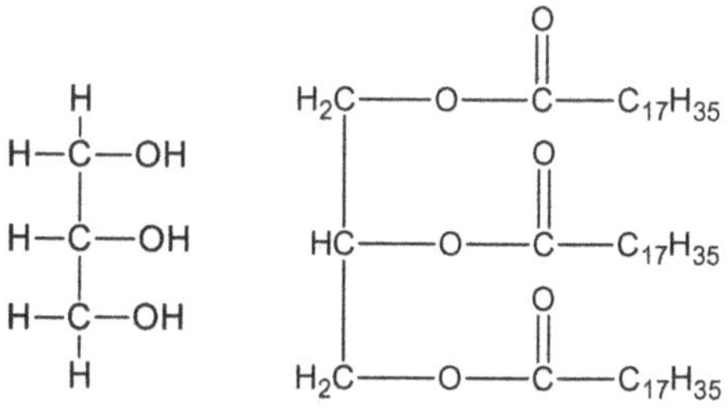

Fig. 11.4. The chemical formulae of (a) glycerol and (b) tristearin.

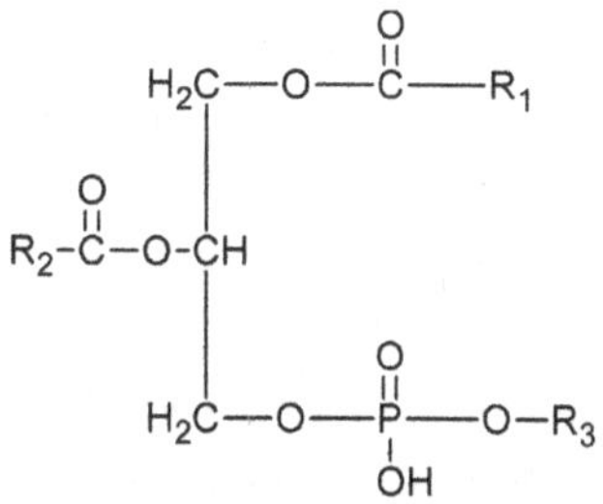

Fig. 11.5. The structure of a phospholipid. R_1 and R_2 are hydrocarbon chains derived from fatty acids esterified with glycerol; R_3 is an organic molecular fragment.

11.3.3. *Phospholipids*

We may visualise the structure of a phospholipid as a glyceride (see Fig. 11.4b), in which one of the fatty acids is substituted by a phosphate group to form the ester, as shown in Fig. 11.5. R_1 and R_2 are chains originated from some of the fatty acids and R_3 is a simple organic molecule.

The molecule which more often enters the composition of the phospholipid as the R_3 fragment is *choline* (see Fig. 11.6). Choline is a component of the Vitamin B complex. Also, in the form of the ester acetylcholine, it is the main chemical messenger in the transmission of signals between neurons. This last subject we shall discuss in more detail in section 11.5.

Phospholipids are a major component of cell membranes, together with proteins.

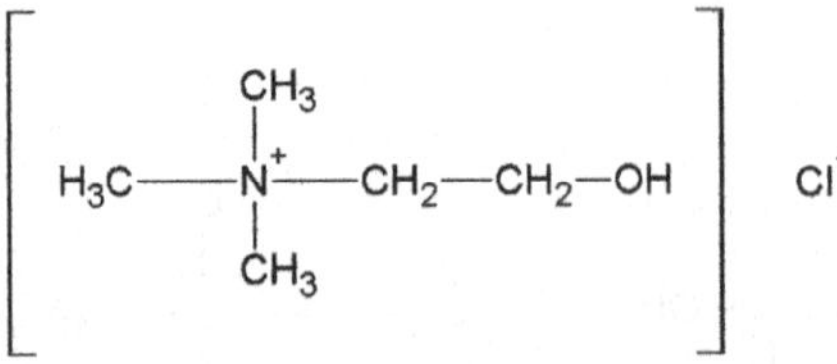

$$H_3C\!-\!\!\!\underset{\underset{CH_3}{|}}{\overset{\overset{CH_3}{|}}{N^+}}\!\!\!-\!CH_2\!-\!CH_2\!-\!OH \qquad Cl^-$$

Fig. 11.6.   The chemical structure of choline chloride.

11.4. Enzymes

In Chapter 9 we explained that a catalyst is an atom, molecule or surface which will increase the rate of a chemical reaction. At the end of the process, the catalyst should be recovered in its initial conditions (of course this is ideal, in practice a catalyst will become exhausted after long periods of use). We may say that *enzymes* play the role of catalysts in biological systems.

Enzymes are proteins and their catalytic function is specific for a single chemical reaction in a given biological system. The enzyme, say, E, acts on another molecule (S) denominated *substrate,* either to break it, as C –> A + B or to put together two molecules to form a new one, say A' + B' –> C'. The specificity of the enzyme is the result of its particular geometry. The enzyme and the substrate have to get together as stereo-specifically as a key and a lock. In Fig. 11.7 we show a very basic representation of the key-lock coupling of two molecules.

The substrate is another molecule present in the particular biological medium: another protein, sugar, lipid, etc.

Many enzymes require the presence of a metal ion in the reaction site to be able to act. Some of these ions are quite common, such as Na^+, K^+ or Zn^{2+}. Others are rather "exotic" metals, which are rarely thought of as associated with living bodies. Examples are elements such as copper, vanadium or molybdenum. Such is the reason we find minute quantities of these elements in commercial vitamin and mineral supplements sold in drug stores.

Incidentally, some enzymes require the cooperation of vitamins to be able to accomplish their function, mainly molecules found in the vitamin B complex.

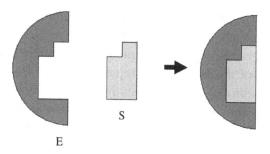

Fig. 11.7. The substrate, S, enters the cavity of the enzyme, E.

Most of us are familiar with the name of the enzyme *pepsin*. Pepsin is found in the stomach of mammals and breaks proteins into smaller fragments, eventually into amino acids, which may enter the blood stream.

11.5. The Transmission of Information Through the Nervous Network

If inadvertently we put a hand close to a pot of boiling water, the information *pain* will travel to the brain and the brain will give the order to the hand, arm and trunk muscles: "take your hand off". At the same time, we are aware that the reaction is not instantaneous. In this case the nervous information or *signal* travels back and forth at a speed of up to some 100 m/second. Had the brain needed to *think*, the reaction would be slightly slower.

We shall discuss the main steps of this complex mechanism, with beautiful chemistry and physics involved. The cells responsible for nervous conduction are called neurons; the interconnection between neurons in our brain is at the basis of the thinking mind. Human brains contain some 10^{11} neurons; each neuron in the brain is connected to seven other neurons. Neurons are also distributed all over the body.

Neurons are cells with the form of an axis and variable length; some neurons in the human body may have a length of 1.5 m. This axis receives the name of *axon*. At both ends of the axon, there are organelles called *dendrites*. The nervous signal flows through the axon and then to the dendrite of the next cell.

When a nervous impulse is generated, for instance by an aggression to a part of the body (heat, injury), the impulse is transported through the neuronal system. Two distinct mechanisms are put at work to carry the information: a transmission mechanism which could be denominated as electrical and another which has a chemical nature. We describe these two processes below.

First we refer to the transmission of the signal along the axon. Both inside and outside the neuron cell, sodium (Na^+) and potassium (K^+) ions are present, dissolved in the intra- and extra-cellular water. The two phases are separated by the cell membrane. The signal propagates as sodium ions are pumped out of the cell, while potassium ions enter the cell. This polarisation wave displaces at a speed which may reach 100 m/second, as we said before. This signal is of the *on* or *off* type, there is no influence on the intensity of the wave whatsoever. Instead, the higher or lower intensity of a stimulus is a result of the number of neurons activated. Of course, the propagation of the polarisation wave is at the expense of energy provided by the body's biochemistry system.

Thus, the transmission occurs through a potential change, analogous but not identical to the electric current in a cable. Still, the axon ought to be protected by an insulator; this role is accomplished by *myelin*, mainly formed by lipids. The disease multiple sclerosis is due to the lack of the myelin sheath, since then the electric potential is dissipated and conduction of the signal fails.

When the electric signal reaches the end of the axon, it stimulates the dendrite. Between the dendrite of one nervous cell and the next there is water, of course. The junction between the two neurons is called the *synapse*. The terminal dendrite liberates a chemical messenger, which flows to the next neuron and there stimulates a new electrical impulse. And so on, until the impulse reaches the brain or, vice versa, to a given terminal cell. Figure 11.8 shows a schematic representation of a synapse between two neurons.

Let us pay attention to the chemical species nature has selected as messengers. It is interesting that these neurotransmitters are so

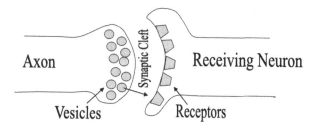

Fig. 11.8. Schematic representation of a neural synapse.

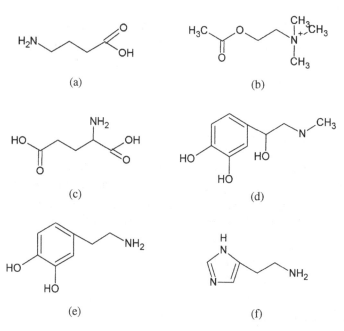

Fig. 11.9. Some common neurotransmitters: (a) gamma-amino butyric acid (GABA); (b) acetylcholine; (c) glutamic acid; (d) adrenaline or epinephrine; (e) dopamine; (f) histamine.

specific in their role (some types of information require unique types of molecules to function) and, at the same time, embrace a vast range of molecular structures. In Fig. 11.9 we list some common neurotransmitters and show the chemical structures.

Molecules in the body may have more than a single function. For instance, in Fig. 11.9 adrenaline is also a *hormone*. Hormone is a messenger, in the sense that it is produced by some cells of the body and affects other cells, eventually distant from the site the hormone was produced.

Glutamic acid is well known as the mono-sodium salt (MSG), in which the proton (H^+) of one of the carboxylic groups (–COOH) is replaced by the sodium ion (–COONa). It is at least odd that this neurotransmitter is used also as food additive in countries in which this is allowed.

Curiously, diatomic toxic gases also act as neurotransmitters. It is so in the case of carbon monoxide (CO) and nitrogen oxide (NO). Obviously, the amount of these molecules in the synapses is extremely low, but sufficient for the transmission of the neural information.

11.6. The Mechanism of Vision

We shall focus our discussion here on the mechanism of vision in vertebrates. The anatomy, function and components of eyes might have been "invented" several times during the evolution of different species. The eye is such a complex and well-designed machine that Charles Darwin himself (who was recently celebrated for the bi-centenary of his birth; England 1809–1882) in his book *On the Origin of Species,* sought to convince that this was a possible result of natural selection. For all species, visual pigments contain retinal, derived from vitamin A, and the family of proteins called opsins.

Visual pigments are located in the retina of the eye and are the molecules sensitive to light. The opsin protein, through the amino acid lysine is combined with the aldehyde retinal to form a Schiff base. Figure 11.10 shows the molecular structures of Vitamin A (the chemical name is retinol, an alcohol), the aldehyde retinal, and the Schiff base formed between retinal and lysine, one of the amino acids composing the opsin. The chemical formula of lysine is shown in Fig. 11.2c.

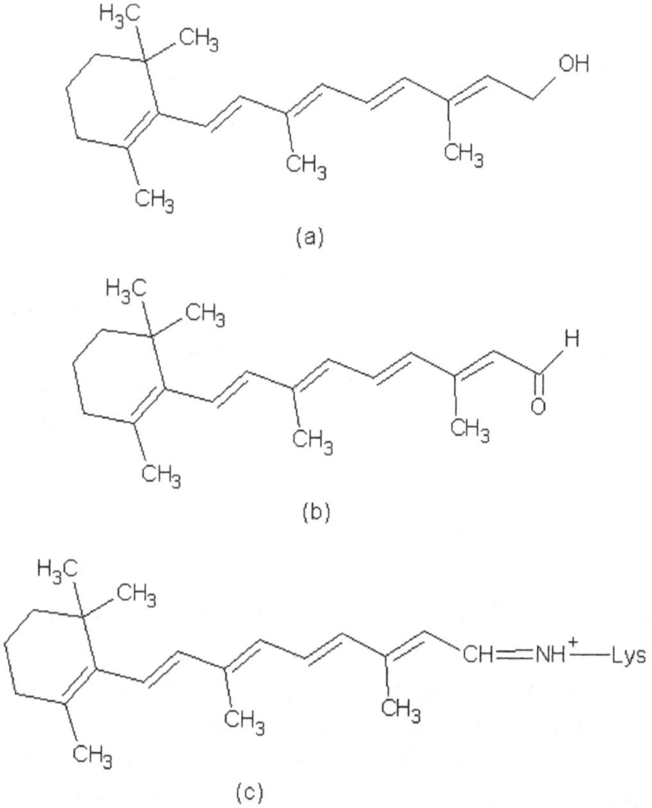

(a)

(b)

(c)

Fig. 11.10. The molecular structures of (a) retinol (Vitamin A); (b) retinal (the aldehyde derived from Vitamin A in the all-*trans* form); (c) Schiff base formed between the amino group of lysine (Lys) and retinal (as the protonated 11-*cis* isomer).

The scheme below shows the transformation of an alcohol, $R-CH_2OH$, to the corresponding aldehyde, $R-CHO$, with the loss of a water molecule:

$$R-CH_2OH \rightarrow R-CHO + H_2O$$

This reaction occurs in the metabolic processing of Vitamin A before the synthesis of the visual pigment, called *rhodopsin*.

The visual pigment rhodopsin is thus formed by the Schiff base resulting from the reaction between retinal (R–CHO) and lysine (H_2N–Lys):

$$R-CHO + H_2N-Lys \rightarrow RCH=N-Lys + H_2O$$

A proton from the surrounding water is attached to the nitrogen atom as is illustrated in Fig. 11.10c; the same Fig. 11.10b shows the retinal component in an all-*trans* ("linear") conformation. The all-*trans* retinol is transformed by biochemical processes into the bent (11-*cis*) form, which is the normal form in retina in the absence of light.

The absorption of a photon initiates a series of biophysical and biochemical events, one of the first being the back isomerisation from 11-*cis* into all-*trans*. These events lead ultimately to the generation of an electrical impulse which carries the information of light reaching the retina to the brain.

Two types of cells are present in retina, denominated according to their form: *rods* and *cones*. In the case of rod-type cells, light is detected as described in the preceding paragraph with extraordinary efficiency; it is believed that a single photon is detectable by the eye of mammals.

The cone-type cells, while not as efficient as rods, have the ability to distinguish colors. In the case of humans, three types of cones exist, each with a different protein part (opsin) of the visual pigment and absorbing respectively in blue, green and yellow regions of the visible light spectrum. With the superimposing of such information, the brain creates the enormous number of colors and nuances of colors we sense. Let us mention that for certain individuals several types of color blindness may occur.

11.7. Heredity

Heredity or genetics is a vast field, embracing chemistry, psychology and ultimately, religion. In this context, it is interesting to remember that Charles R. Darwin, other than Medicine, was also a student of Theology. Darwin published several books and articles. Those with probably greater impact include the book *On the Origin of Species,* published in 1859 and the book on the influence of sex in human descent, from

1871. Darwin's observations and deductions are astonishingly manifest in modern science. It is fair to remember that the also British scientist Alfred Russell Wallace (1823–1913) had simultaneously similar ideas as Darwin and that they exchanged correspondence.

To what extent are humans the result of their genetic inheritance, in contrast to habitat, social and family adaptation and, how much room is left for free will? Genetics itself, has, at least, two components, say the atavic inheritance which, for instance, makes us feel pain when touching a hot pot (some individuals do not have the sensation of pain) or the mistrust of somebody standing behind us. And then there are characteristics we owe to our parents and ancestors, such as the colour of eyes or the probability of the growth of some kinds of cancer.

The reader can imagine that the chemistry and biochemistry pertinent to this magnificent endeavor are immensely complex. For the purpose of this text, we shall describe the background chemical structures of deoxyribonucleic acid (DNA). DNA molecules carry the genetic information. The basic units of DNA are complex molecules (have the generic name of nucleotides) as depicted in Fig. 11.11.

The central compound is a five-member ring formed by four carbon atoms and one oxygen atom. The –C–O–C– segment is characteristic of an organic *ether* function. There are also two alcoholic groups: a hydroxyl (–OH) group linked to the C_2 atom or the C_3 atom in the ring and an external –C–CH_2OH group as substitutent on atom 4. The chemical name of this molecule is deoxyribofuranose and it belongs to the family of *sugars*. Glucose and fructose are common sugar components in daily life; the molecule formed by the association of the two latter

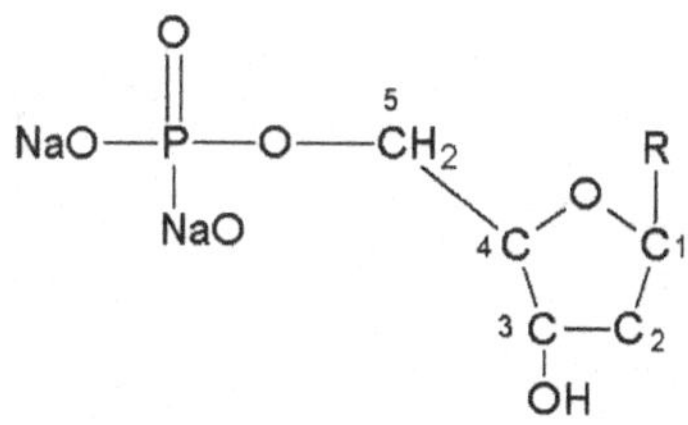

Fig. 11.11. The basic structure of a nucleotide. R represents diverse nitrogen hetero-cycles which are described below. The conventional numbering of the central ring atoms is also shown.

compounds is the common table sugar, called by chemists as sucrose or saccharose.

Phosphoric acid, in the form of phosphate forms an *ester* with the last alcohol group. The other two valences of phosphorus are used by a sodium (Na^+) ion or are ionized as ($-O^-$).

The symbol $-R$ in Fig. 11.11 represents a variety of conjugated nitrogenated heterocycles derived from pyrimidine shown in Fig. 11.12a. The pyrimidine bases cytosine (Fig. 11.12b), uracil (c) and thymine (d) are the main components of nucleotides. These molecules play the role of the substitutent $-R$, as illustrated in Fig. 11.11. The linkage between the five-membered ring and $-R$ is through the N atom indicated in Fig. 11.12a.

Other than the pyrimidine bases shown in Fig. 11.12, bi-cyclic molecules may participate in the DNA structure, as the nitrogenated cycles adenine and guanine, shown in Fig. 11.13.

Nucleotides are linked to each other in the DNA molecule through an ester bond between the phosphate fragment of one and a hydroxyl group of the next as shown in Fig. 11.14.

R_1 and R_2 represent two different or equal nitrogenated bases.

In living beings, the DNA consists of a large number of nucleotide units, in the order of millions of them. Each nucleotide carries one of

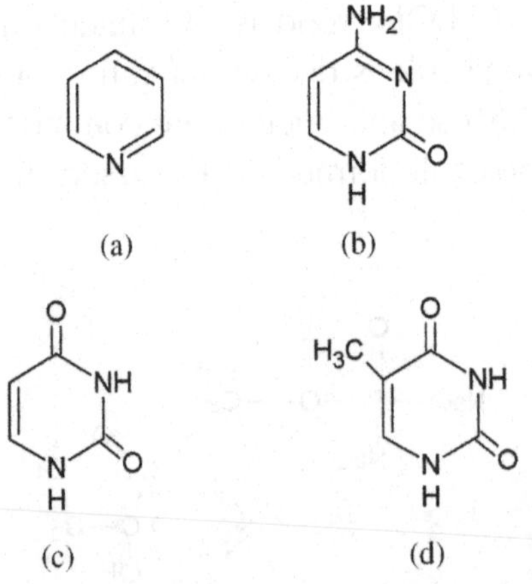

Fig. 11.12. Pyrimidine (a) and the derived bases cytosine (b), uracil (c) and thymine (d). The numbering of the atoms of the ring is defined by choosing the N atom in pyrimidine as number 1.

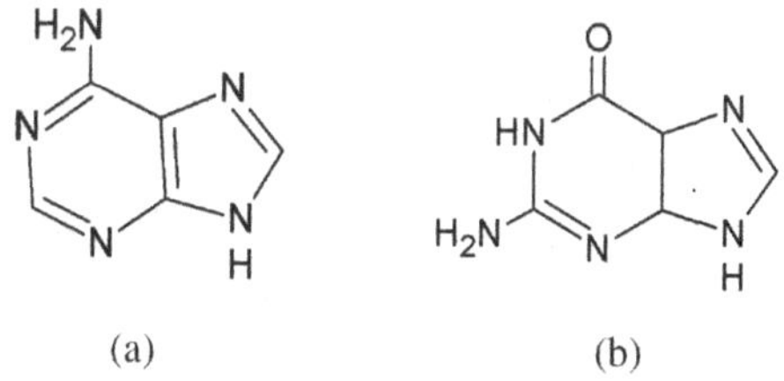

Fig. 11.13. DNA constituting bases adenine (a) and guanine (b).

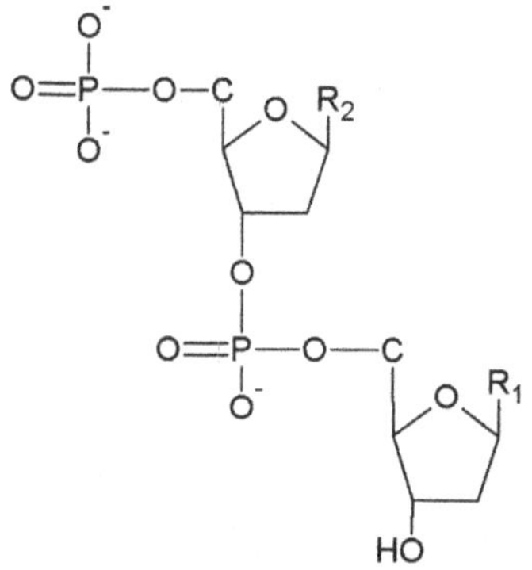

Fig. 11.14. Illustration of the bonding between two nucleotides.

the nitrogenated cyclic bases adenine (A), cytosine (C), guanine (G) or thymine (T) and the number of possible sequences of these four bases is significantly large. These sequences carry the genetic information.

DNA naturally occurs in a double-stranded form, like two ribbons, with nucleotides on each strand complementary to each other. The more technical term refers to the two chains of nucleotides as the double helix. Watson (James Watson, USA, 1928) and Crick (Francis Crick, England, 1916–EUA, 2004) communicated this structure for DNA in 1953. They deduced this structure analysing X-ray crystallographic results obtained by Rosalind Franklin and Maurice Wilkins at King's College in London, UK.

When the two chains of nucleotides split, each one regenerates a new "ribbon", identical to the departed partner, thus two identical copies of the initial DNA molecule are formed; that is how inheritance is accomplished.

In the sequence, DNA controls the order of the amino acids in proteins, determining structures and functions for subsequent generations.

Let us end this chapter on heredity and this book with a note of hope. Chapter 10 conveyed the sentiment that humanity lacks the ability to cope with man-made or natural environmental disasters. We may say in our defense that the Universe is believed to have existed for some 14 billion years and we started leaving the caves a mere ten thousand years ago. Still, something ought to be done. Either we create technologies to maintain our way of life or something similar to it, or we will have to change it. Yet, if we go to extremes and implement radical ecological restrictions, more people are likely to end up starving. Finding a balance therefore, will be the greatest challenge for humanity and the scientific endeavour.

Final Remarks

While writing this book we constantly touched on social, economic and technological issues which are current challenges for humanity. We avoided entering too deep into these matters so as not to lose our main chemistry focus. In these Final Remarks we address broader issues not always connected directly with chemistry.

Let us hope that in the case of technology the extrapolation to some years ahead (say, no more than two decades) is somewhat safer than in economy and finances. Of course, some spectacular scientific breakthrough may arise, but its wide technological application would not be instantaneous.

Nowadays, two energy sources are dominating — fossil fuels and hydroelectricity. At the same time we are also witnessing the rise in a number of alternatives.

Governments, businesses and the people are becoming increasingly environmentally conscious. We are living in days of environmental dramas. There is the oil spill which hurt Louisiana and is now menacing Mississippi, Alabama and Florida. Yet another oil spill has occurred in the Yellow Sea near the Chinese city of Dalian. And nobody seems to care much about the five decades-old petroleum leak in the Niger Delta in Nigeria.

There are however, some energy sources we could consider as limitless. The solar energy which reaches the earth in a couple of years is more than all the fossil fuel reserves combined. Indeed, we do not know how to master all that energy; neither do we wish to do so. But the planet benefited from that energy for a long, long time, and life did as well. Man is starting to harness just a fraction of that power. This is a fast growing alternative, being incorporated into all kinds of devices.

The energy captive in the atomic nucleus is also a source of almost limitless energy. The number of nuclear plants across the world is also increasing. However, this does have the drawback of radioactive waste production.

There is a growing public conscience in the wealthiest nations for the need of a concerted effort from governments, businesses and citizens. People are building a consensus claiming "green" products, methods, engines, homes and, ultimately, a "green" planet. So decision makers at all levels are being pushed (willing or not) to start to behave with more respect towards nature. The degree of "greenness" of proposed solutions will now weigh heavily in the decision for or against their acceptance.

The case of atomic power is a very advanced and sophisticated example of the possibility of "cleaning" a procedure. The atomic power plants use the fission of atoms as the source of energy (Chapter 10). These reactions produce radioactive isotopes which are a major problem for storing and discarding. There is an international project to create a fusion plant. The energies so created would be enormous and, in principle, have less contaminating hazards as compared to the present plants. Clearly we are referring to the normal functioning of such future installations. Accidents are a separate matter.

A main culprit of air pollution is gasoline. Producing gasoline without sulfur is one way to diminish sulfur oxides and sulfur acids in the engine exhausts. Another improvement has been the installation of catalysers to produce the complete combustion of gasoline, so as to eliminate a large part of the toxic gas, carbon monoxide.

Ethanol producers are researching technologies for obtaining the product from ever less valuable materials, such as bio-mass, with little intrinsic value otherwise.

For some years we can expect a popularization of the flex-cars, meaning flexible fuel vehicles. Brazil will certainly push for that solution for some time, which is not too difficult to implement and decreases the CO_2 gas expelled into the atmosphere. Brazil has

9.3 million of such vehicles, followed by the US with 9 million. Canada and some countries in Europe have also a significant number of flex cars.

Curiously, car sellers in the US do not inform the buyer he has bought a car with such capability, so the ethanol option is under-utilised. In contrast, some 6 million cars in Brazil solely employ ethanol regularly.

Fuels have yet another use as natural gas for some 11 million vehicles, mainly in Pakistan, Argentina, Iran and Brazil.

Simultaneously, we have a race of alternative solutions, some of them hybrids between two energy sources. Main hybrids are based on fuels and electricity. Thus they are still fuel dependent. Some such cars are today being produced and in operation. There are 1.6 million of such vehicles active in the US. The major manufacturer is Japan, with some 2.3 millions sold.

Let us now comment on some of the options for fuel-free cars. We have the electric car based on a set of batteries, which demand space and are heavy. Solar cars are ideal, but still need an electrical complement for practical use. Hydrogen cars may use the gas merely as a combustible or with fuel cells generating electricity to power the engine. In either case the product is water, making this a very "clean" option.

One can foresee that fuels — fossil or renewable — will continue in the automobile industry for at least for two decades, if not more.

Which are the best candidates for the future? Maybe the hydrogen cars due to their high degree of "greenness". But nothing is perfectly "clean". It is not merely water pouring from the exhaust. There will be catalysers, hydrogen storage materials and so on.

Recycling appears as an excellent idea and it is hard to believe somebody would oppose it. But even this issue is controversial! Its opponents claim that recycled products are often more expensive than new products and the jobs linked to recycling are less valued than those for new products. In some instances this is true. What needs to be valued, even more than the financial aspects, is the culture which is captivating populations: Reduce, Reuse, Recycle. This is a precious reaction of the society and governments have to preserve it. It may even be true that chemicals used to recycle paper are contaminants. Still we ought to continue recycling paper.

There are variable degrees of recycling. For instance, a used computer may serve merely to extract some gold from it. Or, in a different approach, some pieces in perfect function could be used for new computers. It is likely that extracting and installing the good components may be more expensive than assembling only new components.

Other economic and cultural aspects may play a role. For instance, Brazilians are the second or third largest beer consumers in the world. And there is a mass of people for recovering beer and pop cans, and making a living from it. So Brazil recycles 80% of the aluminum cans.

Thus, it may be the case that "green" is less efficient and economical at times. But we ought to prefer green. Not with the fanatic postures of the eco-radicals, but with sensitive choices.

We also need to beware of "fake green". There is much buzz in the media on natural products, organic food and phytotherapeutics. There are indeed many benefits in some of these products such as the valerian (*Valeriana officinalis*) used by our great-grandmothers. But the axiom that what is natural is good is a myth. There are many toxic plants. The reader may indeed know that Socrates was killed with hemlock (*cicuta*), a natural plant. Natural products and phytotherapeutics intended for medicinal purposes should be submitted to the same rigorous tests for activity and side-effects as are drugs from the pharmaceutical industry.

At this point one can mention the debate on legalization of marijuana (*Cannabis sativa*) for medical purpose. There are some 4000 chemicals in marijuana and indeed some may be beneficial. For instance, one of us, in research work,[1] has found derivatives of cannabin to be more powerful as an analgesic than morphine. But the chemical or pharmaceutical approach is to isolate such analgesic molecules and test them, like any medical drug.

One can make the following analogy. Let's imagine that, among the 5000 chemicals in ordinary tobacco one active molecule against some disease is found. Would you encourage cigarette smoking for that reason?

Sometimes, appropriate legislations may be useful too. There was an active commerce of illegally exported wood of Amazonian trees for Europe. When there is a demand, there is always a supply. This is true for even worse things, like trafficking of drugs and people. Now the European Union (EU) has established the need for the Brazilian authority's certification for exported wood. This measure, if respected, may save millions of trees in the Amazon.

There is room for some straight chemistry here. It so happens that universities — the chemistry departments mainly — are significant consumers of chemical reagents. This is the case in both the teaching and research laboratories. Among the large variety of chemicals, a significant proportion (in terms of volume of material) consists of solvents in which many experiments are performed. Solvents include water, toluene, benzene, acetone, ethanol, and many, many others. Once used for one experiment, these solvents become contaminated and are useless for further experiments.

Let us describe some approaches to overcome the problem of these used solvents. First, minimization of experiments has gained followers, i.e., performing the same experiments with minimal amounts of reagents.

We shall narrate the treatment given to the chemical residues in our own São Carlos Campus, University of São Paulo. Some 15 years ago the Laboratory for Chemical Residues was created. It was an expensive decision for the University, involving buildings (one of which is for storage with high aeration and reduced risk of electrical discharges), equipment and staff (a PhD in Chemistry and a Security Engineer, and several working fellowships for students).

In this installation a large part of the chemicals, mainly the solvents, are recycled to adequate purity so as to be reused by the originating laboratories. Toxic residues are transformed into as inert materials as possible; a simple example is sulfuric acid, extremely toxic, which is neutralized with some sodium base, becoming the rather non-menacing sodium sulfate. At the end, there is a very modest residue that needs to be incinerated or deposited with sanitary waste.

The last subject we wish to consider is the availability of fresh water for agriculture and of drinking or potable water for humans and animals. The planet is a closed system for water with evaporation, transpiration and precipitation involving land and seas, with water either as liquid, vapor, snow or ice. In the more developed countries there is an adequate supply of drinking water per capita. Contrarily, as expected, the number of people with no access to clean drinking water in the so called "developing" countries is increasing. Not to mention the lack of appropriate sewage treatment systems.

The three R's — Reduce, Reuse and Recycle — are especially relevant for water. Systems of conservation could be implemented for cities, communities or homes. There is a large variety of alternatives. Water from kitchen sinks and showers could be used for the discharge of toilets. Such kinds of water could also be directed to treatment for reusing. Rainwater could be saved and used for various purposes, even for drinking, depending on the degree of contamination of the atmosphere the rainfall traversed.

Interview with Professor Rezende (Environmental Chemistry)

What You Always Wanted to Know About Environment and Never Knew Whom to Ask

Interviewee:
Professor Maria Olímpia de Oliveira Rezende (MOOR), Ph.D, D.Sc.
Head of the Laboratory of Environmental Chemistry
Instituto de Química de São Carlos
University of São Paulo

Interviewer:
Professor Milan Trsic (MT)
Instituto de Química de São Carlos
University of São Paulo

MT: Could you define Environmental Chemistry (EC) in non-technical terms?

MOOR: Chemistry is the science which studies why and how chemical reactions occur. So the chemist can diagnoste problems and suggest solutions at the microscopic level. Nonetheless, in the case of EC one needs a global view. Our students are trained in chemistry focused in the preservation of the environment. The student needs to understand the physical, chemical and biological consequences,

157

mainly anthropic, in the planet processes. Displacement and storage of materials may have xenobiotic (chemical compound not found normally in the human body) influence on the life of human beings. Thus, EC considers the elements, compounds or substances whose displacement, transport, and final destination influence and are influenced by environmental conditions. At all times the environmental conditions are to be correlated with public health and economics, with the binomium development and sustainability always present.

MT: Which specific areas of EC are you interested in researching?

MOOR: Soils, waters and humic substances (HS).

In the case of soils, we evaluate the fertility, contamination with pesticides and toxic elements and work for the repair of impacted soils.

For waters we work in the processing of waste water, either for discarding in rivers or for drinking use, in both cases obeying Legal Norms. We also work in the detection of chemical contaminants.

Regarding HS one must know that the organic matter existing in soils, peat, sediments and natural waters is a complex material, evolving with time under the influence of physical, chemical and biological factors. The resulting material after the transformations may be classified in two large groups: *the non-humic substances*, mainly proteins, amino acids, polysaccharides, fatty acids and other chemical compounds, and the *humic substances* (HS), a heterogeneous mixture of various molecules with high molecular weights and different functional groups, being responsible for diverse natural processes.

HS have a relevant role in the environment, mainly due to the variety of functional groups and the widespread presence of the HS in the earth crust. The negative charges, depending on the concentration of H^+ ions, allow the HS to participate in most of the chemical reactions in soils. That is, HS participate in the biogeochemical cycles on the earth surface. The poly functional character of the HS makes them capable of complexing diverse metallic ions. The HS also absorb organic pollutants as pesticides, for instance. Our research on HS focuses on their environmental importance, their interaction with pesticides and various elements. We also study vermicomposting (this means earthworms metabolize the soil material) for the production of HS and for the recovery of impacted soils.

MT: Which are the main international scientific meetings related to your area?

MOOR: In my specific area, the most important is the biannual meeting of the International Humic Substances Society (IHSS). In between these meetings, we hold a Brazilian congress. Brazilian participation in the IHSS is quite significant and the current President of the Society is a Brazilian from EMBRAPA (Brazilian Government institution for research and support for agriculture and cattle raising).

MT: At the present moment two energy sources are prevailing: fossil fuels and hydroelectricity. A number of alternatives are competing namely, nuclear, solar, tidal, hydrogen, wind power, renewable fuels, and so on. Which of them in your view is winning the race so far?

MOOR: Nuclear plants seem to be leading.

MT: Do you believe this a good solution?

MOOR: No. Disposal of radioactive waste has not been resolved. There is no way man can change the half-life (half-life is the time the mass of a given isotope is reduced to half of its initial mass) of radioactive isotopes.

MT: There is a multinational project to build a nuclear plant using fusion instead of fission to generate energy. What are your views on this?

MOOR: One may expect less radioactive contamination, *a priori*. I shall not comment further since it is too far from my specialization.

MT: Do you have a preference or prediction for the best sources of energy humanity should rely on?

MOOR: There is no single answer. There are various good options depending on the location. Let me give two extreme examples: you can bet on solar energy in Brazil but you would not choose tidal power in Switzerland.

MT: You explained that Brazilian scientists in the environmental field have a respected place in the international scientific community. Also, Brazil has the experience of the Itaipu dam, the first ethanol program, the Amazon forest, 14% of the fresh water of the planet, millions of cars moved by ethanol and created the flexible car technology. Nevertheless, my opinion with respect to the world political climate meetings is that Brazil's voice is seldom heard. Would you say that the media is minimizing Brazil's environmental achievements, or the most influential countries do not listen to Brazilian terms with enough interest or that Brazilian representatives simply have too little punch and presence on the international stage?

MOOR: I believe the latter is true. Brazilians tend to overestimate what comes from abroad.

MT: At times one hears the claim that sugar cane is pushing the Amazon forest. Is there any truth in that?

MOOR: There is no sugar cane growing in the Amazon region whatsoever. Sugar cane is grown mainly in the State of São Paulo and in some states in the northeast of Brazil.

MT: As a scientist in this area, do you agree that there is global warming in course?

MOOR: We do not have values of earth temperatures beyond some 100 years back, but we do have evidence that there is a warming in course. For instance, the melting of the polar ice is a fact.

MT: There have always been temperature fluctuations. Do you think that in this case there is a human contribution?

MOOR: We can blame man indeed, mainly for the increase of carbon dioxide in the atmosphere, very significant in the last century. Carbon dioxide is one of the gases of the greenhouse effect, together with water vapor, methane, nitrous oxide, and ozone. Increase of methane gas is also a man-made effect.

MT: How do these gases affect the temperature?

MOOR: There has always been a global warming process, by the absorption of infrared radiation (let us say heat) in our atmosphere. This process was accomplished by water, carbon dioxide, naturally. What we are talking about is actually an extra global warming due to the *increase* in the amount of carbon dioxide and methane.

MT: Let us turn our attention to water — drinking water or potable water. Millions of humans have no access to clean drinking water, mainly due to underdevelopment. How do you think humanity should react?

MOOR: This is a case in which the three R's are mandatory: Reduce, Reuse and Recycle. This may be implemented for homes, for communities or for cities. Bath residual water may be diverted for the flush of toilets. In the end, all water has to be recycled. Japan is apparently the country with more advanced solutions for water conservation.

MT: Do you think that some environmentalist groups defend positions which are not in the interest of humanity?

MOOR: I definitely disagree with the propositions of banning pesticides and transgenic food. Radical positions do not help. Each case has to be evaluated carefully. We need to know the cost of health, the cost of hydric resources, and the cost of well being. We need a discipline to study Environmental Economy.

Brief Curriculum Vitae of Professor Rezende

Graduated in Chemistry at the University of Sâo Paulo (USP) in 1980. Obtained her Ph.D. from the same institution in 1987. Associate Professor at the Institute of Chemistry, Campus São Carlos, USP since 1989. Visiting Professor at the Oklahoma State University, 1992–1994. More than 90 published articles and three books written. Represents the USP in the Federal Department of Environment in the field of security of chemicals.

Notes

Chapter 2

1. Using the former denomination for positive and negative charges, *antimatter* is formed by nuclei in which the anti-protons have negative charge surrounded by positrons (positively charged particles with the mass of electrons).

Chapter 3

1. This refers to normal natural processes; appropriate manipulation may lead to the formation of molecules with energy higher than that of their components (see Chapter 9).

Chapter 7

1. This is the historical presentation of the Born and Oppenheimer Approximation. In fact, the movement of particles of equal mass can also be separated, as is the case for electrons.
2. The exact solution, E, requires solving the equations for E_{ee}; this correction, although relatively small — say some 2% of the total energy — demands heavy computational cost in time and memory. The effort to calculate this correction is not a mere theoretical refinement. Often the purpose of our calculation is the difference between the total energies of two systems (for instance, between an atom in the ground state and the same atom in the excited state; or the difference in energy between a system and its ionic species). These differences are accessible to experimental determinations and may have the same order of magnitude as E_{ee}.

3. Linear combination is an expression used in mathematics to denominate the combination of variables in which all the variables have exponent 1(normally one writes $x^1 \equiv x$). This is certainly a particular case; for instance, $C_1 x^2 + C_2 y^2 = 10$ is not *linear* since the variables x and y have exponent 2.

4. Linus Pauling was awarded the Nobel Prize in Chemistry in 1954 and the Nobel Peace Prize in 1962, the latter for his tenacious efforts against nuclear weapons.

Chapter 8

1. There are subtle chemical differences though, which may be at the origin of separation methods — for the enrichment of uranium-235 for instance.

Chapter 9

1. As in the case of polymers, the use of the word *infinite* for the number of monomers of a polymer means actually: very large (some 300 to several thousands) but is not a synonym of the term *infinite* as a mathematical concept. In the present case, the term *universe* has also a dual nature. The universe of an experiment we may perform at the bench in the laboratory takes little account of events occurring in a galaxy, say 10^4 light years away.

2. The symbol *d* in mathematics means *differential* increment. Would one be interested in the whole evolution of the system, it would be necessary to integrate over the path.

3. It is outrageous that at the same time millions of people are starving. This is far more a geopolitical than a chemical issue.

Chapter 10

1. One has to be cautious; man is searching for life outside of the earth, and it may be, that if it exists, it will be of a very different nature to the familiar forms. Even in our planet, life does exist in completely dark habitats. For instance, in a gold mine in

South Africa, 2.8 km below the earth's surface, the bacterium *Desulforudis audaxviator* was discovered. The ultimate source of energy for this species seems to be radioactive decay of uranium.
2. As this book goes to print, a huge oil spill in the Gulf of Mexico is devastating the east coast of North America.

Chapter 11

1. In May 2010, Dr. J. Craig Venter (J. Craig Venter Institute, USA) announced the creation of a "synthetic cell". This breakthrough is creating a storm of scientific and political controversies.
2. Contrarily, in the case of *periodic* polymers, units are repeated, as in polythiazyl.

Final Remarks

1. Saulo L. da Silva, Agnaldo Arroio, Albérico B. F. da Silva, and Milan Trsic, *A correlation between geometric features and analgesic activity for a series of cannabinoid compounds*, J. Mol. Structure **441** (1998) 97–100.

Further Reading

We have selected a few books, separated for various purposes. There are options for the formal learning of chemistry, some challenging incursions into knowledge, and some positions on where we are concerning the environment.

Textbooks on Chemistry

Atkins P., Jones L. (2009). *Chemical Principles: The Quest for Insight.* W.H. Freeman, New York.

Brady J., Senese F. (2008). *Chemistry: The Study of Matter & Its Changes.* Student Study Guide, Hoboken.

Snyder C. H. (1995). *The Extraordinary Chemistry of Ordinary Things.* John Wiley, New York.

Book published in 1935, when modern Chemistry and Physics were starting

Pauling L., Wilson E.B. (1935). *Introduction to Quantum Mechanics. With Applications to Chemistry.* McGraw-Hill, New York and London.

Challenges in Science

Gilmore R. (2001). *The Wizard of Quarks: A Fantasy of Particle Physics.* Springer, New York.

Gilmore R. (1955). *Alice in Quantumland: An Allegory of Quantum Physics.* Copernicus, New York.

Horgan J. (1996). *The End of Science,* Addison-Wesley, Reading, Massachusetts.

Arntz W., Chasse B., Vicente M. (2007). *What The Bleep Do We Know?* Health Communications, Deerfield Beach, Florida.

Climate

Adams J. (1995). *Risk* Routledge-Taylor & Francis, London and New York.
Ponte L. (1976). *The Cooling: Has the Next Ice Age Already Begun?* Prentice-Hall, New Jersey.
Leggett J.K. (ed.) (1990). *Global Warming: The Greenpeace Report.* Oxford University Press, Oxford.
Hulme M. (2009). *Why We Disagree About Climate Change: Understanding Controversy, Inaction and Opportunity.* Cambridge University Press, Cambridge.

Chemistry for Beginners

Robertson W.C. (2007). *Chemistry Basics: Stop Faking It! Finally Understanding Science So You Can Teach It,* The NSTA Learning Center, Arlington.
Lagowski J.J. (2004). *Chemistry: Foundations and Applications.* Macmillan Reference USA, Woodbridge.
Williams L. (2003). *Chemistry Demystified a Self-Teaching Guide.* McGraw-Hill Professional, New York.
Rosenberg J.L., Epstein L.M. (1996). *Schaum's Outline of College Chemistry,* 8 edition, McGraw-Hill, New York.

Web Sites for Self Learning

Organic Chemistry

http://www.organic-chemistry.org/
http://www.neok12.com/Organic-Chemistry.htm

Analytical Chemistry

http://ull.chemistry.uakron.edu/analytical/index.html

Biological Chemistry

http://www.biology.arizona.edu/

General Chemistry

http://antoine.frostburg.edu/chem/senese/101/index.shtml

Inorganic Chemistry

http://www.chemguide.co.uk/inorgmenu.html

Physical Chemistry

http://www.chemistry-blog.com/category/physical-chemistry/

Chemistry Experiments

http://bizarrelabs.com/
http://www.csiro.au/resources/ChemistryActivities.html
http://chem.lapeer.org/Chem1Docs/
http://www2.uni-siegen.de/~pci/versuche/english/versuche.html
http://www.chem.umn.edu/outreach/Demos.html
http://www.sciencebuddies.org/science-fair-projects/recommender_interest_
 area.php?ia=Chem
http://www.woodrow.org/teachers/chemistry/institutes/1986/

APPENDIX I

Spherical Co-ordinates

$$r = \sqrt{x^2 + y^2 + z^2}$$

$$\phi = arctan\frac{y}{x}$$

$$\theta = arctan\frac{\sqrt{x^2 + y^2}}{z}$$

$x = r\sin\theta\cos\phi$

$y = r\sin\theta\sin\phi$

$z = r\cos\theta$

Periodic Table of the Elements

Group →	1	2	3	4	5	6	7	8	9	10	11	12	13	14	15	16	17	18
	H 1 1.0079																	He 2 4.0026
	Li 3 6.941	Be 4 9.0122											B 5 10.811	C 6 12.011	N 7 14.007	O 8 15.999	F 9 18.998	Ne 10 20.180
	Na 11 22.990	Mg 12 24.305											Al 13 26.982	Si 14 28.086	P 15 30.974	S 16 32.065	Cl 17 35.453	Ar 18 39.948
	K 19 39.098	Ca 20 40.078	Sc 21 44.956	Ti 22 47.867	V 23 50.942	Cr 24 51.996	Mn 25 54.938	Fe 26 55.845	Co 27 58.933	Ni 28 58.693	Cu 29 63.546	Zn 30 65.39	Ga 31 69.723	Ge 32 72.61	As 33 74.922	Se 34 78.96	Br 35 79.904	Kr 36 83.80
	Rb 37 85.468	Sr 38 87.62	Y 39 88.906	Zr 40 91.224	Nb 41 92.906	Mo 42 95.94	Tc 43 [98]	Ru 44 101.07	Rh 45 102.91	Pd 46 106.42	Ag 47 107.87	Cd 48 112.41	In 49 114.82	Sn 50 118.71	Sb 51 121.76	Te 52 127.60	I 53 126.90	Xe 54 131.29
	Cs 55 132.91	Ba 56 137.33	Lu 71 174.97	Hf 72 178.49	Ta 73 180.95	W 74 183.84	Re 75 186.21	Os 76 190.23	Ir 77 192.22	Pt 78 195.08	Au 79 196.97	Hg 80 200.59	Tl 81 204.38	Pb 82 207.2	Bi 83 208.98	Po 84 [209]	At 85 [210]	Rn 86 [222]
	Fr 87 [223]	Ra 88 [226]	Lr 103 [262]	Rf 104 [261]	Db 105 [262]	Sg 106 [266]	Bh 107 [264]	Hs 108 [269]	Mt 109 [268]	Uun 110 [271]	Uuu 111 [272]	Uub 112 [277]		Uuq 114 [289]				

* 57–70

** 89–102

* Lanthanide series

La 57 138.91	Ce 58 140.12	Pr 59 140.91	Nd 60 144.24	Pm 61 [145]	Sm 62 150.36	Eu 63 151.96	Gd 64 157.25	Tb 65 158.93	Dy 66 162.50	Ho 67 164.93	Er 68 167.26	Tm 69 168.93	Yb 70 173.04

** Actinide series

Ac 89 [227]	Th 90 232.04	Pa 91 231.04	U 92 238.03	Np 93 [237]	Pu 94 [244]	Am 95 [243]	Cm 96 [247]	Bk 97 [247]	Cf 98 [251]	Es 99 [252]	Fm 100 [257]	Md 101 [258]	No 102 [259]

Constants and Conversion Factors

Mass (SI = kg)

	kg (kilograma)	lb (Pound)	oz (ounce)
kg	1	2,20462	35,27399
lb	0,453592	1	15,99999
oz	0,0283495	0,0625	1

1t = 10^3 kg ; 1 kg = 10^3 g = 10^6 mg

Distance (SI = m)

	m (metro)	ft (foot)	in (inch)	mi (mille)
M	1	3,2808	39,37	$6,2137 \times 10^{-4}$
Ft	0,3048	1	12,00	$1,8939 \times 10^{-4}$
In	0,02540	0,08333	1	$1,5783 \times 10^{-5}$
Mi	1609,3440	5280,00	63360,0	1

1Å = 10^{-10}m; 1 m = 10^2 cm = 10^3 mm

Volume (SI = L)

	L	ft³	in³	gal (USA)	qt (USA)
L	1	$3,531467 \times 10^{-2}$	61,023759	0,2641721	1,0566881
ft³	28,316847	1	1728,0004	7,4805231	29,922077
in³	$1,638706 \times 10^{-2}$	$5,787035 \times 10^{-4}$	1	$4,329005 \times 10^{-3}$	$1,73160 \times 10^{-2}$
gal (USA)	3,785410	0,1336804	230,99995	1	3,9999979
qt (USA)	0,946353	$3,342014 \times 10^{-2}$	57,750017	0,2500001	1

1 L = 1 dm³ = 10^{-3} m³; 1 mL = 1 cm³ = 10^{-6} m³

Pressure (SI = Pa)

	Pa(Pascal) $(N.m^{-2})$	Atm	bar	Torr (mmHg)	psi $(lbf.in^{-2})$
Pa	1	$9,86923 \times 10^{-6}$	$1,0 \times 10^{-5}$	$7,500617 \times 10^{-3}$	$1,45 \times 10^{-4}$
Atm	$1,01325 \times 10^5$	1	1,01325	760	14,696
bar	$1,0 \times 10^5$	0,9869233	1	750,0617	14,504
Torr	133,3224	$1,315789 \times 10^{-3}$	$1,333224 \times 10^{-3}$	1	0,0193
Psi	$6,895 \times 10^3$	0,06805	0,06895	51,715	1

Energy (SI = J)

	J (N.m)	Kcal	eV	kWh	BTU
J	1	$2,390057 \times 10^{-4}$	$6,241512 \times 10^{18}$	$2,777778 \times 10^{-7}$	$9,478134 \times 10^{-4}$
kcal	4184	1	$2,611448 \times 10^{22}$	$1,62222 \times 10^{-3}$	3,965651
eV	$1,602176 \times 10^{-19}$	$3,829293 \times 10^{-26}$	1	$4,450489 \times 10^{-26}$	$1,518564 \times 10^{-23}$
kWh	$3,600 \times 10^6$	860,4207	$2,246944 \times 10^{25}$	1	3412,128
BTU	1055,06	0,2521654	$6,585169 \times 10^{21}$	$2,930722 \times 10^{-4}$	1

System cgs: $1 \text{ erg} = 10^{-7} \text{ J}$

Power (SI = W)

	W (J.s)	$kcal.h^{-1}$	hp	$BTU.min^{-1}$
W	1	0,8604207	$1,341022 \times 10^{-3}$	$5,68181 \times 10^{-2}$
$kcal.h^{-1}$	1,162222	1	$1,558565 \times 10^{-3}$	$6,61375 \times 10^{-2}$
hp	745,7	641,6157	1	42,408
$BTU.min^{-1}$	17,6	15,12	$2,36 \times 10^{-2}$	1

Temperature (SI = K)

	Kelvin (K)	Celcius (°C)	Fahrenheit (°F)	Rankine (°R)
K	1	$T_c = T_K - 273,15$	$T_F = \dfrac{9}{5}T_K - 459,67$	$T_R = \dfrac{9}{5}T_K$
°C	$T_K = T_C + 273,15$	1	$T_F = \dfrac{9}{5}T_C + 32$	$T_R = \dfrac{9}{5}(T_C + 273,15)$
°F	$T_K = \dfrac{5}{9}(T_F + 459,67)$	$T_C = \dfrac{5}{9}(T_F - 32)$	1	$T_R = T_F + 459,67$
°R	$T_K = \dfrac{5}{9}T_R$	$T_C = \dfrac{5}{9}T_R - 273,15$	$T_F = T_R - 459,67$	1

Prefixes:

10^{-12}	pico (p)	10^1	deca (da)
10^{-9}	nano (n)	10^2	hecto (h)
10^{-6}	micro (m)	10^3	Kilo (k)
10^{-3}	milli (m)	10^6	mega (M)
10^{-2}	centi (c)	10^9	giga (G)
10^{-1}	deci (d)	10^{12}	tera (T)

Constant

Constant	Symbol	Value
Avogrado`s Number	N_A	$6{,}02214 \times 10^{23}$ mol^{-1}
Molar Gas Constant	R	$8{,}31451 \times 10^{-2}$ L bar K^{-1} mol^{-1}
		$8{,}31451$ J K^{-1} mol^{-1}
		$8{,}31451$ L kPa K^{-1} mol^{-1}
		$8{,}20578 \times 10^{-2}$ L atm K^{-1} mol^{-1}
		$62{,}3639$ L Torr K^{-1} mol^{-1}
Boltzmann Constant	k	$1{,}3806503 \times 10^{-23}$ J K^{-1}
Speed of Light (Vacuum)	c_o	$2{,}9979458 \times 10^8$ m s^{-1}
Planck Constant	h	$6{,}62606876 \times 10^{-34}$ J s
Faraday Constant	F	$9{,}64853415 \times 10^4$ C mol^{-1}
Standard gravity	g	$9{,}80665$ m s^{-2}
Charge on a Proton/Electron	e	$1{,}602176462 \times 10^{-19}$ C
Atomic Mass constant	ua	$1{,}66053873 \times 10^{-27}$ kg

The Most Stable Ions of the Common Elements

H^+	Hg^{2+}	Sn^{4+}
H_3O^+	Mn^{2+}	Pb^{4+}
NH_4^+	Cr^{2+}	Pt^{4+}
Li^+	Fe^{2+}	Ti^{4+}
Na^+	Co^{2+}	As^{5+}
K^+	Ni^{2+}	Sb^{5+}
Rb^+	Sn^{2+}	F^-
Cs^+	Pb^{2+}	Cl^-
Ag^+	Pt^{2+}	Br^-
Cu^+	Ti^{2+}	I^-
Hg_2^{2+}	Al^{3+}	N_3^-
Au^+	Bi^{3+}	H^-
Be^{2+}	Mn^{3+}	O^{2-}
Mg^{2+}	Cr^{3+}	O_2^{2-}
Ca^{2+}	Fe^{3+}	O_4^{2-}
Sr^{2+}	Co^{3+}	S^{2-}
Ba^{2+}	Ni^{3+}	C_2^{2-}
Ra^{2+}	Au^{3+}	Se^{2-}
Zn^{2+}	As^{3+}	Te^{2-}
Cd^{2+}	Sb^{3+}	N^{3-}
Cu^{2+}	Mn^{4+}	P^{3-}
		C^{4-}

The Most Stable Ions of the
Common Elements

Isotopes of the Elements and Their Relative Abundance

Main Isotopes with % Occurrence

H	**N**	**Al**	**K**
1–99.98%	14–99.63%	27–100%	39–93.3%
2–0.02%	15–0.37%		41–6.7%
		Si	
He	**O**	28–92.2%	**Ca**
4–100%	16–99.8%	29–4.7%	40–96.9%
	18–0.2%	30–3.1%	42–0.7%
Li			44–2.1%
6–7.5%	**F**	**P**	
7–92.5%	19–100%	31–100%	**Sc**
			45–100%
Be	**Ne**	**S**	
9–100%	20–90.5%	32–95.0%	**Ti**
	21–0.3%	33–0.8%	46–8.0%
B	22–9.3%	34–4.2%	47–7.3%
10–19.9%			48–73.8%
11–80.1%	**Na**	**Cl**	49–5.5%
	23–100%	35–75.8%	50–5.4%
C		37–24.2%	
12–98.63%	**Mg**		**V**
13–1.1%	24–79.0%	**Ar**	50–0.3%
	25–10.0%	36–0.3%	51–99.7%
	26–11.0%	40–99.6%	

Cr
50–4.4%
52–83.8%
53–9.5%
54–2.4%

Mn
55–100%

Fe
54–5.9%
56–91.7%
57–2.1%

Co
59–100%

Ni
58–68.1%
60–26.2%
61–1.1%
62–3.6%
64–0.9%

Cu
63–69.2%
65–30.8%

Zn
64–48.6%
66–27.9%
67–4.1%
68–18.8%

Ga
69–60.1%
71–39.9%

Ge
70–21.2%
72–27.7%
73–7.7%
74–35.9%

As
75–100%

Se
74–0.9%
76–9.4%
77–7.6%
78–23.8%
80–49.6%
82–8.7%

Br
79–50.7%
81–49.3%

Kr
80–2.3%
82–11.6%
83–11.5%
84–57.0%
86–17.3%

Rb
85–72.2%
87–27.8%

Sr
86–9.9%
87–7.0%
88–82.6%

Y
89–100%

Zr
90–51.5%
91–11.2%
92–17.2%
94–17.4%
96–2.8%

Nb
93–100%

Mo
92–14.8%
94–9.3%
95–15.9%
96–16.7%
97–9.6%
98–24.1%
100–9.6%

Ru
96–5.5%
98–1.9%
99–12.7%
100–12.6%
101–17.1%
102–31.6%
104–18.6%

Rh
103–100%

Pd
102–1.02%
104–11.1%

105–22.3%
106–27.3%
108–26.5%
110–11.7%

Ag
107–51.8%
109–48.2%

Cd
106–0.9%
108–0.9%
110–12.5%
111–12.8%
112–24.1%
113–12.2%
114–28.7%
116–7.5%

In
113–4.3%
115–95.7%

Sn
112–1.0%
116–14.5%
117–7.7%
118–24.2%
119–8.6%
120–32.6%
122–4.6%
124–5.8%

Sb
121–57.4%
123–42.6%

Te
122–2.6%
123–0.9%
124–4.8%
125–7.1%
126–18.9%
128–31.7%
130–33.9%

I
127–100%

Xe
128–1.9%
129–26.4%
130–4.1%
131–21.2%
132–26.9%
134–10.4%
136–8.9%

Cs
133–100%

Ba
134–2.4%
135–6.6%
136–7.9%
137–11.2%
138–71.7%

La
138–0.1%
139–99.9%

Ce
140–88.4%
142–11.1%

Pr
141–100%

Nd
142–27.1%
143–12.2%
144–23.8%
145–8.5%
146–17.2%
148–5.8%
150–5.6%

Sm
144–3.1%
147–15.0%
148–11.3%
149–13.8%
150–7.4%
152–26.7%
154–22.7%

Eu
151–47.8%
153–52.2%

Gd
154–2.2%
155–14.8%
156–20.5%
157–15.7%
158–24.8%
160–21.9%

Tb
159–100%

Dy
160–2.3%
161–18.9%
162–25.5%
163–24.9%
164–28.2%

Ho
165–100%

Er
164–1.6%
166–33.6%
167–23.0%
168–26.8%
170–14.9%

Tm
169–100%

Yb
170–3.1%
171–14.3
172–21.9%
173–16.1%
174–31.8%
176–12.7%

Lu
175–97.4%
176–2.6%

Hf
176–5.2%
177–18.6%
178–27.3%
179–13.6%
180–35.1%

Ta
180–0.01%
181–99.99%

W
182–26.3%
183–14.3%
184–30.7%
186–28.6%

Re
185–37.4%
187–62.6%

Os
186–1.6%
187–1.6%
188–13.3%
189–16.1%
190–26.4%
192–41.0%

Ir
191–37.3%
193–62.7%

Pt
194–32.9%
195–33.8%

196–25.3%

198–7.2%

Au

197–100%

Hg

198–10.0%

199–16.9%

200–23.1%

201–13.2%

202–29.9%

204–6.9%

Tl

203–29.5%

205–70.5%

Pb

204–1.4%

206–24.1%

207–22.1%

208–52.4%

Bi

209–100%

Po

209–100%

At

210–100%

Rn

222–100%

Fr

223–100%

Ra

226–100%

Ac

227–100%

Th

232–100%

Pa

231–100%

U

235–0.7%

238–99.3%

Index

absolute temperature scale 104
acetic acid 42, 136, 137
acetone 44, 136
acetylcholine 139, 143
acetylene 26, 38, 54, 86, 87
acid rain 126
activated complex 115, 118
activation energy 115, 117–120
additives 119
adenine 148, 149
adenosine biphosphate 113
adenosine triphosphate 113
ADP 113
adrenaline 143, 144
Aeolian energy 122
agriculture 56, 57, 97
agrochemicals 127
alanine 134, 135
algae 40, 101, 127, 129
aliphatic hydrocarbons 34
alkaline 12, 45, 57, 88, 123
alkaline elements 12
alkaloids 45
alkanes 34–37
allotropes 114
alloy 21
alpha particles 92
amides 45

amino acids 46, 47, 134–136, 141, 144, 150
amino group 45, 135, 145
ammonia 23, 45
amphetamines 39
anaesthetic 40
angular quantum number 7, 10, 72
aniline 45
anion 12, 13, 53, 54
anode 128
anthracene 27
antibiotic 42
anti-detonant 125
aquifers 56, 57
aromatic bond 27
atmosphere 16, 17, 25, 80, 97, 122, 125–127, 130
atom-atom distances 79
atomic charges 79
atomic functions 6, 7, 10
atomic nucleus 1, 3, 66, 91, 94
atomic orbitals 9–11, 17, 18, 24, 26, 39, 71–73, 75, 77
atomic structure 1, 4, 6, 65, 66
ATP 113
axon 141, 142

bacteria 42, 101, 119
band gap 52, 53
band theory 51
battery 122, 127, 129, 131
benzaldehyde 38
benzene 27–29, 34, 38, 39, 53,
    81, 83, 86, 87, 89, 90, 136,
    137
beverages 41, 42, 119
blood stream 50, 141
Bohr model 4
Bohr radius 4
boiling point 31, 35, 57, 104
Boltzmann law 83
bond energy 31
bond orders 79, 86, 87
brain 108, 134, 141, 142, 146
butyric acid 143

C_6H_6 39, 81, 136
calcium 22, 38, 50, 134
calorie 106
camphor 44
car exhaust 119
carbohydrates 47, 48
carbon dioxide 23, 25, 97, 116,
    126, 127, 137
carbon monoxide 17, 77, 125,
    144
carbonate 22
carbonic acid 45
carboxylic group 135
carcinogenic 40
Cartesian coordinates 10
catalyst 118, 119, 125, 140
cathode 128
cation 13
cell membrane 142
cellulose 47, 127

Celsius temperature scale 104
chain reaction 38, 98
chemical affinity 115
chemical bonds 1, 16, 20, 87,
    115
chemical energy 97, 101, 124
chemical kinetics 114
chemical transformations 2, 101,
    102
Chernobyl 124
chloride 17, 49, 88, 137, 140
chlorine 12, 13, 22, 40, 49, 50,
    88
chloroform 29, 40, 136
chlorophyll 101, 127
choline 139, 140
chromium 21, 51
citric acid 46
clean energy 17
cleaning products 41
climate 122, 131
clusters 57–59, 61, 77, 84
CO 17, 42, 44, 45, 72, 77, 78,
    125, 144
coal 39, 121, 126
collision 115, 117
combustion 2, 17, 35, 50, 116,
    119, 121, 122, 124–127, 137
computers 6, 63, 77, 82
concentration 88, 116–118
conducting polymer 122
cone-type cells 146
cooling 121
copper 21, 22, 49–51, 54, 55,
    140
covalent bond 22–24, 28, 31
cracking 36–38
crystal 22, 50, 88, 114
crystalline structure 21

cyclohexanone 44
cysteine 134, 135
cytosine 148, 149

dam 121, 122
degenerate functions 72
dendrite 142
density 1, 7, 18, 19, 21, 59, 75,
    86, 88, 91, 125
deoxyribofuranose 147
diatomic molecules 16, 17
diesel 36, 127
diesel oil 36
dimer 57, 58
dipole moment 28–30, 60, 61
discrete values 64
distillation 35, 124, 126
DNA 39, 95, 112, 147–150
dopamine 45, 143
dopants 54
double bond 23, 26, 28, 37, 42,
    138
double helix 149
drugs 39, 67, 136

ecological 150
electric conductivity 51, 54, 55
electric conductors 22, 24, 51,
    54, 55
electric field 59
electrical energy 98, 120, 122
electrical impulse 142, 146
electricity 50, 120–122, 128–130
electrolysis 129
electrolyte 128
electron 1, 5–9, 13, 17–26, 28,
    65, 69–73, 77, 86, 88, 92, 93,
    95, 104
electron density 1, 18, 19, 86, 88

electron shells 22
electronegativity 21, 23, 28, 58,
    88, 89
electronic configuration 8, 16, 25,
    77
electronic energy 5, 6
electronic structure 1, 3, 10, 15,
    66, 70, 72, 91
electrostatic interaction 21
elementary particles 1, 3, 63, 65
endothermic 85, 110, 115, 116
energy levels 5, 6, 9, 13, 51, 52,
    72
energy sources 2, 56, 97, 121,
    122
enthalpy 85, 109–111, 119
entropy 107–109, 111, 114
environment 2, 102, 103
enzyme 31, 140, 141
equilibrium constant 112
equilibrium distance 19, 20, 80, 82
esterification 138
esters 43, 44
ethane 35, 80, 81, 84, 86, 87
ethanol 37, 41, 42, 121, 126,
    127, 130, 131, 136–138
ethane 25, 26, 37
ethene 25, 26, 37
ethers 42
ethyl ether 42
ethylene 26, 37, 38, 86, 87
ethylic alcohol 41
excited states 8–10, 79, 95
exothermic 85, 110, 115, 116,
    119
explosion 98, 114

Fahrenheit temperature scale 104
fat 106

fatty acids 48, 137–139
fauna 122
fermentation 42, 126
Fermi level 52
fertilizers 131
flora 122
fluoride 28, 58
fluorine 28, 40, 58
food 43, 44, 47, 49, 50, 55, 56,
     97, 101, 106, 114, 119, 121,
     131, 144
formaldehyde 42, 44
formol 44
fossil energy 35
fossil fuels 124
fresh water 56
fructose 47, 147
fuel 2, 35, 42, 121–124, 126–129,
     131
fuel cell 128, 129

GABA 143
gamma rays 95
gas phase 83, 85, 110, 115
gasohol 127
gasoline 27, 36, 37, 120, 121,
     124, 125, 127, 131
Gaussian type functions 11
gene 136
genetic information 2, 95, 147,
     149
genetic mutation 39
genetics 24, 146, 147
Gibbs free energy 109, 111, 114
global warming 126, 130
glucose 47, 50, 112, 113, 127,
     147
glutamic acid 143, 144
glyceraldehydes 47

glycine 134, 135
gravity 4, 123
ground state 8, 25, 72, 74, 75, 80,
     151
guanine 148, 149

habitat 105, 109, 137, 147
haemoglobin 50, 51, 125
halogenated derivatives 40
halogens 12, 35, 57, 88
heat 25, 35, 37, 50, 56, 85, 86,
     98, 102–104, 106, 108–110,
     121, 122, 137, 142
heat capacity 56
heating 8, 33, 35, 36, 39, 121,
     122, 129
Heisenberg's Uncertainty Principle
     19
helium 16, 99
heredity 146, 150
Hess' law 109
hetero-cycles 147
histamine 143
HOMO 73–75
hormone 37, 44, 144
hybridization 24–27
hydrate 129
hydroelectric plants 122, 124
hydrogen atom 1, 3–7, 23, 30, 43,
     69, 71, 129
hydrogen bond 30, 31, 42, 58
hydrogen cyanide 23
hydrogen fuel cells 121
hydrogen sulphide 46
hydrogenation 48
hydroxyl group 42, 45, 148

independent electron
     approximation 70–72

inorganic molecules 1, 49, 84
insecticides 97
insulator 53, 142
intermolecular forces 28, 32, 57
internal energy 105, 106, 109
inter-nuclear distance 17
ionic bond 21, 22, 24, 88
ionization potential 65
ions 10, 12, 13, 20–22, 24, 50,
   128, 140, 142
iron 21, 22, 50
iso-electronic 20
isolated system 102
isomerise 48
isomers 84, 124
isooctane 36
isotopes 95–97

Kelvin temperature scale 104
ketones 44, 47
kinetic energy 18–20, 70, 118,
   120

lead 27, 43, 70, 92, 98, 106, 124,
   125, 146, 151
leptons 63, 93
linear combination 73, 75, 152
linoleic acid 138
lipids 47, 134, 136, 137, 142
lithium 22, 131
living beings 1, 2, 15, 33, 39, 40,
   44, 46, 49, 50, 101, 105, 119,
   123, 133, 134, 137, 148
London dispersion forces 32
LUMO 73–75
lysine 134, 135, 144–146

macromolecules 134
magnetic field 10, 11, 72

magnetic quantum number 10
many-electron atoms 1, 65
mechanical energy 122
melting point 57
mercury 50
metabolism 25, 45, 50, 97, 112
metallic bond 21
methane 23, 25, 26, 35, 73–76,
   116, 124
methanol 42, 130, 136
metric 104, 105
micro-organism 43
molecular orbital 17, 18, 67, 69,
   72–75, 77, 86, 88
molecular structure 1, 13, 15, 83
morphine 39
MSG 144
myelin 142

NaCl 12, 17, 22, 49, 88
naphthalene 27, 39
natural gas 121
nervous conduction 2, 141
nervous impulse 142
nervous signal 142
neuron 141, 142
neurotransmitters 142–144
neutron 92, 94, 123
n-heptane 36, 37
nicotine 45
niobium 131
nitrate 22
nitrogen 23, 30, 49, 53, 88, 97,
   125, 144, 146, 147
nitrogen oxide 144
noble gases 12, 15, 16
nuclear charge 11, 12, 20, 92
nuclear energy 97, 123, 124
nuclear power plant 123

nuclear reactions 97
nuclear waste 124
nucleotide 147, 148
nylon 44

oceans 40, 56, 130
octane 36, 37, 124, 125
octane rating 37
octet rule 16, 22
odor 44
oil 35–38, 124, 127, 130, 131,
    137, 153
oleic acid 48, 138
open system 102
opsin 144, 146
organic molecules 33
organic solvents 34, 136
ozone 40, 80

palmitic acid 137
pasteurise 42
Pauli's Exclusion Principle 51, 72,
    74, 94
pepsin 141
peptides 136
perfume 41, 42, 44, 46
peroxide 37, 82, 83
pesticide 40
petrol 35, 36, 39, 121, 124–127
phenylalanine 134, 135
π bond 26
phosphate 113, 134, 139, 148
phospholipids 137, 139
photoelectric effect 65
photon 93, 146
photosynthesis 101, 118, 122,
    127, 129
Plank's constant 64

plastic 26
platinum 128, 131
polar bond 28
pollution 38, 121, 130, 131
polyethylene 26, 37, 38
polypeptide 47
polythiazil 54, 55
positrons 95, 151
potential energy 18–20, 70
pressure 38, 103, 105, 109–111,
    115, 118
principal quantum number 4, 8,
    10, 72
products 27, 36, 37, 41, 43, 45,
    48, 85, 111, 112, 115–117,
    119, 126
propane 84, 110
propanoic acid 138
propene 37
propylene 37
protein 24, 51, 107, 135, 136,
    140, 144, 146
proton 5, 57, 69, 92, 94, 144, 146
pyrimidine 148

quantisation 4, 64
quantum chemistry 1, 65–67, 69,
    77, 79
quantum mechanics 1, 19, 63–66,
    114
quarks 63, 94

radical 37, 38, 121, 150
radioactivity 93, 94, 96, 97
rate constant 116, 117
reactants 85, 117
reaction velocity 117
receptor 2, 136

refrigeration 40
reproduction 97, 102
resonance 27
retina 144, 146
retinal 137, 144–146
retinol 144–146
rhodopsin 145, 146
rod-type cells 146
room temperature 22, 27, 31, 47, 137
rotation 80, 84
Rutherford atomic model 93

saccharose 148
saturation 48
Schiff base 144, 145
Schrödinger equation 65
sea tides 123
semiconductor 53
σ bond 26
silver 50, 51, 54
single bond 23, 27, 42
Slater type functions 11
sodium chloride 17, 49, 88, 137
solar cells 122, 129, 131
solar energy 55, 112, 122, 127, 129
solvent 27, 35, 40, 42, 136, 137
spherical coordinates 10
spin 9, 24, 65, 72, 94
standard heat of formation 85
state function 105, 107, 109
stearic acid 48, 137, 138
subatomic particles 63
substrate 140, 141
sucrose 47, 148
sugar 42, 101, 126, 127, 131, 137, 140, 147, 148

sulfanilamide 45
sulfate 22
sulphur hydride 57
sulphur nitrides 53
sun 4, 99, 101, 122
surroundings 102, 103, 108–112
synapse 142, 143
synthesis 33, 37, 44, 67, 86, 111, 117, 145
system 5, 6, 8, 20, 54, 71, 72, 85, 93, 102, 103, 105, 106, 108, 109, 111, 112, 125, 129, 140, 142, 151, 152

temperature 22, 27, 31, 47, 50, 54, 56–59, 83, 84, 102–106, 110, 111, 114, 117, 118, 130, 137
testosterone 44
tetraethyl lead 125
thermal equilibrium 103, 104
thermodynamics 64, 67, 102–105, 107, 114, 118
thermoelectric plants 121
Three Mile Island 124
thymine 148, 149
toluene 38, 39, 90, 136
total energy 16, 19, 20, 70, 79–81, 84, 85, 105, 151
tranquilisers 39
triglycerides 48, 137, 138
triple bond 23, 38

universe 4, 16, 63, 102, 108, 109, 150, 152
uracil 148
uranium 3, 123, 152, 153
urea 33, 45

valence band 52
valence shell 15, 16, 22, 23
van der Waals forces 17, 18, 31
vertebrates 137, 144
vibration 79, 105
vinegar 42, 43
virus 133
vision 2, 137, 144
visual pigments 144

vitamin B 139, 140
vitamins 2, 50, 82, 140

wood 121
work 2, 65, 66, 83, 102, 103,
    106, 108, 109, 111, 118, 121,
    128, 129, 142

xenon 16